Dreams
Betrayed

Dreams
Betrayed
Working in the Technological Age

by

CARLTON ROCHELL
with

Christina Spellman

Lexington Books

D.C. Heath and Company • Lexington, Massachusetts • Toronto

Library of Congress Cataloging-in-Publication Data

Rochell, Carlton C.
Dreams betrayed.

Bibliography: p.
Includes index.
1. Automation—Social aspects—United States.
2. Labor supply—United States—Effect of technological
innovations on.   I. Spellman, Christina.   II. Title.
HD6331.2.U5R63   1987   331.25   85–45019
ISBN 0–669–11105–8 (alk. paper)

Published simultaneously in Canada
Printed in the United States of America
Paperbound International Standard Book Number: 0–669–11105–8
Library of Congress Catalog Card Number: 85–45019

The paper used in this publication meets
the minimum requirements of American National Standard
for Information Sciences—Permanence of Paper
for Printed Library Materials, ANSI Z39.48–1984.
♾ ™

ISBN 0-669-11105-8

87 88 89 90 8 7 6 5 4 3 2 1

To Noëlla Fachinetti
in return for
the title of this book and her support

Washington Square
May 1, 1987

Contents

Acknowledgments

This book grew out of research undertaken at the Villa Serbelloni as writer in residence at the Bellagio Study and Conference Center supported by the Rockefeller Foundation, to which I am most grateful. I also want to thank Ira Mothner who advised as only a friend may—bluntly.

Introduction

I S it reasonable to fear that a machine may be doing your job ten years from now? Could your job just disappear, when robots or some other manifestation of automation take over most of the work in your plant or office?

These are real and serious questions. Estimates of unemployment that will result from high technology have ranged all the way up to fifty percent of today's workforce. But it is now becoming clear that underemployment will be far more devastating than unemployment. Vast numbers of skilled and experienced adults, unable to find a niche in an automated society, will be forced into jobs well below the level of both their aspirations and their abilities: steel plant foremen will scrub floors and executive secretaries will serve Big Macs.

The automated age, so exciting to futurists, computer buffs, and entrepreneurs, is likely to mean problems, uncertainty, and substantial economic loss for many of the rest of us. We don't yet seem to realize this, and the men and women who shape our national policies—government officials, corporate chiefs, labor leaders, even academics—have yet to comprehend the risks of high technology, let alone begin devising a remedy.

Perhaps, like us, they have just been dazzled by the electronic show—for it is indeed dazzling. One can only marvel at the microchip technology and artificial intelligence experiments that produce robots with human-like qualities. But, on a less glamorous level, the computer is revolutionizing the workplace from the assembly line and the clerical department to the analyst's cubicle and the executive suite. Word processing has eroded the traditional distinctions be-

tween secretarial and clerical functions. Numerical control machines, machines programmed to cut metal with more precision than a skilled machinist, have transformed tool and die production. Library catalogs that once housed millions of index cards and occupied thousands of square feet of floor space have been put on-line and reduced to a few disks and an array of terminals.

Changes of similar magnitude are underway in all industries, "dirty" as well as "clean." Newspaper reporters are writing their stories on computers linked directly to the composing room, and computer engineers are about to bring forth a system that can lay out the entire paper. The skilled labor that once was a hallmark of the industry is no longer needed.

Robots, the stars of high tech, are being programmed for sensory awareness and refined forms of discrimination and intelligence. Managers understandably love these machines. They take over dangerous and repetitive jobs; they're tireless and seemingly impervious to such human frailties as daydreaming and the desire for better pay. They are moving into steel mills, auto plants, and fruit orchards, but as they do, white-collar as well as blue-collar jobs disappear.

Even if our leaders *were* more responsive to this threat, they would need to deepen their understanding and broaden their repertoire of ideas to deal with it effectively. More than most aspects of our economy, the impact of high technology is international. As a result, the notion that Americans can solve a complex socioeconomic problem such as technological underemployment on their own has little validity nowadays, and it will have none at all in the future. One reason is the role of multinational corporations, which are among the major producers and users of high-tech innovations. By virtue of their ownerships, they cannot let concern for national or local labor forces dictate worldwide policies.

Many of the issues involved in the introduction of high technology were addressed when General Motors developed its innovative Saturn car project. In July 1985, the company and the United Auto Workers announced an agreement that would have radically altered the position of labor. Indeed, the company described the Saturn plant to be opened at Spring Hill, Tennessee, as a "learning laboratory" where the basis for a far different relationship between labor and management would evolve.

Part of the agreement proposed to change the status of the blue-collar worker: all staff would be salaried rather than paid an hourly wage; there would be only four to six job classifications; and part of everyone's salary would be tied to profit and productivity goals. Production would be organized and coordinated by teams working with computer technology. Significant elements of Japanese management theory would also be adopted: more participative decision making, more communication, less of a hierarchy, and *fewer workers producing more cars.*

By late 1986, however, General Motors had cooled considerably on its plans for the unopened Spring Hill plant. GM's priorities were shifting, and it was clear that in whatever form the Saturn project might materialize, it would not be the revolutionary departure that had initially been announced.

If the bright promise of the Saturn agreement subsequently lost its luster, there was even grimmer news on the job front overall. The *Monthly Labor Review* published a study with unmistakable implications for the future of America's job market. It tracked the paths of 5.1 million displaced workers, most of whom had lost jobs in heavy industries. After three years:

only 30 percent had landed equivalent jobs;

nearly 30 percent had to take a pay cut of 20 percent or more, and for nearly half it was 20 percent;

40 percent had found no job and only 25 percent of these were still looking for work;

prominent in this last group were women and blacks.[1]

In the landmark study *The Impact of Automation on Employment, 1963–2000*, Nobel laureate Wassily Leontief and coauthor Faye Duchin estimate that the rapid onset of high tech will by 2000 require 11.4 percent fewer workers to produce the same combination of goods and services required in 1980. If these projections are correct, there will be sharp reductions in the demand for managers, clerical workers, and other middle-level personnel, as well as corresponding increases in the demand for professionals, laborers, service industry

workers, and craftspeople. The need for managers and clerical work-ers will be reduced by more than 40 percent and for sales workers by almost 18 percent. By contrast, 21.5 percent more professional workers will be required.

In the process of these changes, a new society will begin to take shape, a society represented not by the familiar bell-like curve, but by the vintage Coke bottle; that is, to use instructive oversimplifi-cation, a society with substantially larger upper and lower classes, and a diminishing middle class. This could set in motion a social transformation of immense proportions, with potentially staggering implications for the American standard of living, the American work ethic, and eventually the nation's capacity to govern itself in ways we now take for granted.

Meanwhile, the changes now taking place in U.S. employment are so diverse and disconnected that it is difficult to evaluate them. We may read about 25,000 layoffs at AT&T, 5,000 at General Elec-tric, and 35 at the local newspaper. But we seldom add them up. The Bureau of Labor Statistics estimates that 5.1 million workers were displaced from January 1981 to January 1986. This does not represent a grand total of jobs lost, however, since the bureau defines displaced workers as those who have lost their jobs as a result of plant closings or slack work, or those whose positions have been abolished and who have "significant attachment" to work. Invest-ment bankers may know this—and know where new jobs are being created—but how many other people do?

The National Commission on Jobs and Small Businesses in a 1987 report cites testimony by David Birch of the Massachusetts Institute of Technology to the effect that "each year we lose nearly ten percent of our jobs and our businesses large and small. Just to break even . . . every locality in America must replace half of its jobs every five years."[2]

This is pretty much what has been happening in recent years. But closer analysis shows just what kind of jobs are being lost and what kind are being gained. Heavy losses are occurring in the high-paying manufacturing sector. Since 1980, the United States has lost more than a million production jobs paying an average of thirteen dollars per hour. Simultaneously, the number of lower-paying jobs has increased. Since 1980, 5.5 million new jobs in the service and

retail areas were created, paying an average wage around seven dollars per hour.[3]

Striking changes in employment are occurring because of the proliferation of information technologies. They affect the information-based industries: financial services, banking, law, management counseling, accounting, advertising, publishing, and the various news media. But their impact is far wider, affecting all work that involves the flow of information or data from one unit to another. So, the range includes almost everything that isn't straight production.

The increased number of information workers in New York City provides an interesting profile, but not a typical example. In 1958, the information sector accounted for 35 percent of the city's private employment—41 percent of the city's income. By 1984, it accounted for 55 percent of private employment and 58 percent of income.[4] Growth in this sector of the workforce is not limited to information-based firms, but cuts across the traditional distinctions between goods production and service industries, because management and support staff are now considered information workers.

Growth in the information industries, if considered in isolation, provides a rosy vision of the future. Moreover, the information industries are home ground for reporters, writers, researchers, and academics. What is happening in bankrupt Pennsylvania steel towns is far enough out of view to be often ignored. Besides, these misfortunes can be made to serve as an object lesson of what the nation must guard against rather than as a harbinger of structural change.

To be sure, not everybody who might foresee problems with high technology does see them. Take Charles L. Schultze, who chaired President Carter's Council of Economic Advisers. When interviewed, in the early 1980s, Schultze dismissed technology-produced unemployment as "a bogey—no real problem." Then he reconsidered: "If it *is* a problem," he added, "it has been overstated. In the long run, living standards rise only to the extent productivity rises, and technology can be expected to raise productivity. It will not, however, necessarily raise unemployment. People let out in one place usually are picked up in another. High tech does not, in itself, pose a serious unemployment problem." Asked a similar question, in 1985, President Reagan's Secretary of Labor, William E. Brock, responded: "While one type of worker displacement can result from

a change in technology, technological change in the past has not reduced the overall level of jobs. There have, of course, been particular cases where workers in certain industries and occupations have been hurt by technological change, but it is also true that the aggregate effect has been increased employment and higher living standards."[5]

But the question is not one of employment alone. What about the quality of the work the displaced worker finds? How many former managers or blue-collar craftsworkers, accustomed to drawing self-esteem from their work, will be satisfied with lesser jobs? How will they feel about the loss of income, perks, benefits, and vacation time?

These are not questions Schultze or Brock address. Their optimism is buttressed by the upbeat, consumerist attitudes of the new generation of conservatives—and of the materialistic, eye-on-the-ball yuppies who are their natural allies. The popularity of the personal computer and the overall glamour of a high-tech world have helped to mask the issues that are accumulating in their wake. Why should anyone fear technological advances that make life easier and more fun?

There are many obvious reasons for why we might reconsider our unbridled optimism about rapid technological change. Consider the radical change that automation imposes on some types of work.

Containerization, for example, has reduced the need for longshoremen. In the New York port, their union negotiated a contract with the shipping companies that seemed humane, guaranteeing annual income for displaced workers with ten or more years' seniority. But longshoreman Ronald Bell thinks otherwise. "It was fine for a while," Bell said. "But then the guarantee syndrome set in"—that is, the all-too-human tendency to take the money and stop worrying about getting another job. For Bell and others like him, this induced a feeling of hopelessness and a loss of self-respect. As he put it, "What kind of life is that?"

With a guarantee, longshoremen are still required to show for the "shapeup," the morning call at the union hall where work assignments are made. Since there is usually not enough work for all who report, shifts and gangs are assembled on the basis of lowest seniority. That is, the longshoremen with the least seniority are required to work. Those with the most seniority are called last.

But working longshoremen can make more money than high-seniority men on the guarantee. Some working longshoremen earn as much as $80,000 a year, although $30,000 is closer to the average. The guarantee range is $18,000 to $24,000. So, in some ways, being on the guarantee is equivalent to being bought out. As Bell tells it, at some point the "guarantee syndrome" takes over, making men feel apathetic and no longer in control of their lives.

Interviews with out-of-work middle managers turn up some of the same despair. It is estimated that a job search will take roughly one month for each $10,000 of income. Executives who earned $50,000 to $60,000 report that the stress on families and friends is often crippling. Some resort to therapy; others speak of shock, grief, humiliation, and a sense of financial insecurity. Studies indicate that thirty percent of unemployed middle managers remain so for at least seven months. One study by the American Society for Development and Training found that roughly a third of those over thirty-five wind up in jobs paying less than those they previously held. It takes most of them five years to regain their previous salary levels.

A sense of being downgraded also occurs when workers keep their jobs but are shifted to tasks that require less skill and involvement. One secretary we spoke with in a middle-sized architectural office described the installation of dedicated word-processing machines. First, her relationships with the architects she had worked for changed radically. Now part of a typing pool, she missed the contact and follow-up on projects. More important to her, however, was what she spoke of as "a loss of status." Whereas formerly she had been a member of a project team, now she was just a support service.

Finally, there is the trend toward part-time work. While this appeals to all employers (no fringe benefits, less unproductive time), it has additional advantages for those determined to introduce large-scale automation. With machines available around the clock, people can be hired to run them as the ebb and flow of the workload dictates. The word-processing shop, which churns out copy for all comers, operates in precisely this fashion.

In the banking industry, Citibank—which recognized the enormous costs of paper information transfer in the 1970s—was in the vanguard of automation. At one point, in addition to automating transactions and using robots for mail delivery, the bank employed

computer operators from temporary agencies to do posting at night. One of the obvious advantages of hiring temps is lower overhead costs. Citibank added another advantage when it developed a strategy of hiring from a number of different agencies and playing them against each other by dropping one each month. The result was to be sure of getting the best available part-time workers.[6]

Underlying these vignettes of the larger work world is the specter of structural poverty brought on by sustained unemployment or drastic underemployment. To anyone willing to look, the economic recovery of the Reagan years has left a residual, seemingly irreducible, "unrecovered" core. Buffalo and Detroit and other old industrial cities of the North not only look like besieged areas, they give the feeling of impending doom. Even the Sunbelt, until recently the limitless frontier for the rust-belt refugee, is beginning to show signs of strain. House auctions due to repossession are at an all-time high in Houston. The boom that received so much notice seems to have been fairly short-lived; now some families live in their cars.

Looming, too, is the specter of serious disruption in America's high-charged consumer economy. Part-time workers don't buy a great many goods and services; the unemployed buy less. Robots buy none at all, unless you count maintenance costs. But there is more at risk than just a diminished domestic market. A shrinking middle class, with more rich and more poor, sets the stage for growing class resentment and more crime, and markedly increases the danger of social unrest, riots, and violence.

Journalist Robert Kuttner has put his finger on the process that causes the redistribution of jobs and creates a society where most are at the bottom, fewer in the middle, and a few more go to the top.[7] This shift involves the replacement of primary sector manufacturing jobs by automated equipment. Junior-level management and supervisory jobs are already being automated. Senior-level clerical jobs can easily be automated. By clearing out the middle, responsibility is further centralized at the top; routine, minimally skilled occupations are pushed further to the bottom.

According to this model, the results of automation are also distributed asymmetrically. Corporate automation benefits companies in increased productivity ratios through lower labor costs. The consumer benefits from whatever pass-alongs are involved. The costs devolve onto the displaced people, both in terms of what they are

able to consume and in terms of what they are able to do. Evidence is accumulating that a relatively satisfied middle management is becoming a disgruntled service labor pool as downshifting occurs. The long-range implications are less opportunity for advancement and shorter job ladders as upward mobility becomes more and more difficult.

Hard evidence buttresses these assumptions. In 1984, the Bureau of Labor Statistics estimated that the following occupations would provide the greatest number of new jobs in 1990: secretaries, 700,000; nurses' aides and orderlies, 508,000; janitors, 501,000; sales clerks, 479,000. At the low end of the list were electrical engineers, 115,000, and computer programmers, 112,000.[8] In 1985, two studies, one published in the *Monthly Labor Review* and one by the organization 9to5, indicated that the need for secretarial and clerical workers in 1990 might be less than estimated.[9]

If even the more modest projections of job displacement materialize, government, industry, and labor will be forced to respond with new approaches and policies. The challenge will be to manage and possibly reallocate resources to insure the possibility of economic growth and adequate employment in all regions of the country.

Basic to meeting this challenge is the recognition of how a changing workplace will change our nation and its leadership—in government, industry, labor, and finance. Thus, their own best interests should dictate that today's leaders seize the initiative and respond now to what is happening in the workplace, while there is still time to minimize the social and economic consequences of allowing things to drift.

Any number of factors, however, limit the likelihood of timely and effective action, even though there is growing public perception of the probability of radical change. But political constraints limit what government can do; the short-term profit goals of corporate managers limit what industry and the financial community can do; and organized labor persists in protecting the interests of its present membership with little thought to the future.

1

The Springboard of the Automated Workplace: The Computer

T HE major actor in the high-tech revolution is the computer—miracle itself, maker of miracles in the workplace and in our lives. We may take the computer for granted; we may stand in awe and fear of it as something beyond our understanding; we may welcome it as a benefit the equal of penicillin. But what is this reshaper of our lives, and where did it come from?

The Evolution of the Computer

The computer originated in the laboratory, and its enormous potential was initially limited to scientific and military applications. Historically, the demands of warfare have provided great stimulus for technological development. Funds for the development of computers in the United States were made available from defense budgets. The first decimal computer, ENIAC (Electronic Numerical Integrator and Computer), was a product of World War II ballistics research. Prompting further development was the need for rapid collection and manipulation of statistical information by business. Engineers were also quick to recognize how computers could solve complex calculations and modeling problems, while increased international economic competition created a demand for sophisticated telecommunications and rapid information transfer.[1]

The history of the computer is a tale of ambition and genius, filled with eccentric personalities. The machines that now make mathematical computations at increasingly dazzling speeds can be traced back to the abacus and the counting machines of Pascal and Leibniz. By the mid-nineteenth century, the Englishman Charles Babbage was experimenting with an "analytical engine" that was the ancestor of ENIAC.

Babbage, considered the father of the computer, can be thought of as the prototypical Victorian, and his combination of practical and theoretical problem-solving occurs again and again in the evolution of computer design. Throughout his career, Babbage was concerned with creating calculating "engines." He not only devised methods of calculation for his machines, but also developed the tools needed to build them. He spent his own fortune and enormous sums from the British government on his computing machines, the differential engine and analytical engine. In the end, others built the differential engine, using Babbage's design. But Babbage himself built the more complex analytical engine, with its potential for programming through card-coded instructions.

The machines Babbage developed were sophisticated calculators. Their design shows how well he understood both advanced mathematics and the state of manufacturing technology. Babbage also had a knack for adapting other people's discoveries. For example, he appropriated a system for programming looms developed by the French weaver and textile producer Jacquard to solve the problem of entering information into his computer. Jacquard had developed a system for punching cards with instructions so that a loom could reproduce a design or portrait.

Punched cards were also applied to statistical compilation by Herman Hollerith, an American engineer working in the 1880s. Hollerith combined a card-reading machine with a tabulator to produce the world's first mechanical data processor. This machine proved its worth when it was applied to U.S. Census data. It quickly (for those days) provided information on city size, the ratio of rural to urban population, household composition, and the like. Hollerith was quick to see the advantages of electricity. With it, he produced a calculating system far faster than any of its mechanical predecessors. Tested during the 1880 census, Hollerith's electrical card-counting machines completed the transcription of data on 10,500

Saint Louis people in 72 hours and 27 minutes, and the tabulation in 5 hours and 28 minutes. Its closest competitor, the Pidgen chip machine, took 110 hours and 56 minutes to transcribe the information and 44 hours and 41 minutes to tabulate it.

As an inventor, Hollerith also combined traits that can be found repeatedly in the history of the computer. In his case, these were a clear sense of how to use technology to solve certain types of problems and a desire to see commercial applications of his ideas. Although his machines were expensive, they were indispensable. When a number of tabulation machine producers merged, Hollerith's company became part of the conglomerate that came to dominate the computer industry—the International Business Machines Corporation, or IBM.

The first half of the twentieth century brought the development of increasingly sophisticated calculators—both analog and digital computers. Symbolic logic and binary conversion—the transposition of information into a two-unit format (0 and 1)—were critical elements in the production of the first digital computers, as was the development of more elaborate forms of electronic feedback. Experimentation and development were accelerated by the pressing needs of national defense.

By the standards of previous wars, World War II was as much a contest of technology as military skill and strength. The accelerated development of the digital computer, for example, was the result of the U.S. Army's need for precise firing tables for sophisticated guns. In 1941, the differential analog computer was the state of the art in high-speed calculation. When it proved impractical for the ballistics task, the army established a computing research operation at the University of Pennsylvania's Moore School of Engineering. Working under a high-security classification, Moore researchers, including John Mauchly, J. Presper Eckert, Jr., John Brainard, and Herman Goldstine, completed ENIAC in June 1944. The ENIAC project emerged too late in the war to have much impact on ballistics, but it captured the imagination of John von Neumann, the brilliant mathematician, then working at the Los Alamos Scientific Laboratory. At his urging, ENIAC was applied to Los Alamos calculations for the development of the hydrogen bomb.

ENIAC's achievements stirred rivalries among institutions like the Massachusetts Institute of Technology, the Institute for Advanced Study at Princeton, and the Moore School—all then vying

to produce computers. It also stimulated the scientific community in general, because ENIAC established digital computing as a reliable, quick, and expandable means of computation. ENIAC became the standard from which succeeding developments in computers derived. More versatile and more accurate than the analog computer, it could be programmed to respond to a variety of data needs, including alphabet-based languages. However, even before its completion, ENIAC's designers and most scientists familiar with its development knew that what was needed next was some means of storing memory and a more efficient way of programming instructions.

The postwar period was characterized by intense enthusiasm and competition among engineers and managers who envisioned the enormous potential to be realized from further development. Mauchly and Eckert of the original Moore School team developed BINAC, the Binary Automatic Computer, and UNIVAC, the Universal Automatic Computer. In Britain, the Manchester Mark I was developed by a team that included mathematician Alan Turing, who in 1936 had published what is acknowledged to be one of the most important papers in computer science, "On Computable Numbers."

By the early 1950s, IBM had emerged as the most aggressive commercial developer in the field. While the primary thrust of research was still defense, and other governmental applications came second, the computer's potential for systematic information access and processing was widely recognized. Soon it was accepted that, for any number of accounting or recordkeeping activities, the computer could replace the manual worker.

The technology of computers changed rapidly as the number of their possible uses increased. A major breakthrough came when mechanical tape replaced vacuum tubes for storing memory. Another milestone was MIT's Whirlwind Project, which produced the first "real time" computer. Capable of meeting instant information needs, it could be used for assembly lines and air traffic control. Whirlwind employed multiprocessing—that is, the simultaneous processing of several predetermined programs. It thus pioneered the technology of networking by coordinating and organizing information passed among its different units. Among its most valuable assets was the high speed at which information could be exchanged. As opposed to a single machine, Whirlwind was a system that combined several computers and devices to coordinate information. It also was the

first computer to use magnetic core memory and interactive monitors. In the genealogy of computers, it is the parent of the 1960s minicomputer.

The next technological innovations were the transistor (1947) and the integrated circuit (1959). The transistor replaced the vacuum tube. It made possible solid-state circuits, combining many transistors with fewer components and wired connections. Thus, reliability increased and miniaturization became possible. Integrated circuits came next, as sets of solid-state circuits or semiconductor devices were mounted on a single board or piece of plastic. Integrated circuits not only permitted faster and more reliable information transfer, but they also were easier to maintain and less expensive to mass produce.

In 1971 came the programmable microprocessor in the form of a general-purpose logic chip. By combining memory with a logic chip, a microprocessor can be used in any machine that manipulates information. Its applications range from thermostats to programmable home appliances. Heir to the microprocessor is the microchip, which made possible vast memory storage at low cost and accelerated the development of robotics as well as computational computers. Progressively smaller and more sophisticated chips have made possible progressively more sophisticated applications.

Today, different kinds of machines programmed in different languages can communicate via local area networks (LANs) and intelligent terminals act as gateways to stored data. Combined with telecommunications technology, the chip transforms the terminal into an intelligence device with access to a wide variety of data and services from multiple locations.

By the time the personal computer hit the marketplace in the 1970s, microchips had become capable of storing increasing amounts of memory at less and less cost. The advent of the Apple and, later, the IBM Personal Computer, made Americans in general increasingly conscious of the computer and its uses for them. While this diffusion to the public was taking place, leading engineers were designing superpowered machines with super memories, designed to operate at extraordinary speeds. The technique they developed was multiprocessing, which involves the dense packing of chips into a compact C-shaped cabinet to minimize the time it takes for electrical currents to travel from one part of the machine to another. In the Cray-2, which has the world's largest internal memory capacity (2

billion bytes), 240,000 computer chips are used. The Cray-2 is 40,000 to 50,000 times faster than a personal computer. What this represents in speed alone is summed up by Robert Borchers, associate director of the Lawrence Livermore National Laboratory: "What took a year in 1952 we can now do in a second."

These multiprocessors represent a radical new stage in the development of computer technology. Astrophysicist Larry Smarr has assessed it as the "biggest development in forty years." Multiprocessors also create a new set of expectations for still more improvement. Seymore Cray (who founded Cray Research in 1972) describes the current environment: "As far back as I can remember I was very proud to make factor-of-four improvements from one generation of machines to another. Instead, we're moving forward by factors of ten."[2]

As computer technology evolves, the need for programming languages develops along with it. The programs for ENIAC were basically wiring diagrams, and much of the early program language consisted of flowcharts. With FORTRAN (formula translation), which was developed at IBM between 1954 and 1957, computer programming was opened up to anyone who had a good sense of logic and the persistence to learn. Initially used by the IBM 704 computer, it was soon adapted for use in other IBM machines and sold to other manufacturers. This enhanced the ability of computers to communicate with each other, because they could speak the same language.

As commercial demand expanded and new demands arose, responsive computer languages become more varied: COBOL (common business-oriented language), BASIC (beginner's all-purpose symbolic instruction code), and SPSS (special program for the social sciences). These languages have become progressively easier for the nonspecialist to use. The advent of personal computers has brought "user-friendly" screen instructions for such programs as Wordstar, Wordperfect, and Visi-Calc.

As the computer became more accessible and more broadly useful, its application expanded. During the 1950s, insurance companies and banks as well as research and development-related industries began using computers. With the onset of miniaturization, cost ceased to inhibit use by middle-sized companies. Any operation that could

be broken into a logical sequence with specifiable activities was a prospect for computerization.

The Undeniable Advantages

Whether a mainframe or a mini, the computer is basically a capital-intensive, labor-saving device. It rationalizes task completion. It works well in the current corporate structure, which is based on centralized control of decision making. It also adapts functionally to the needs of the new entrepreneurial corporations. Computer networks enable a company or a group to transcend the conventional limits of time and space. Effective networking allows engineers in West Germany and Detroit to collaborate in the design of cars. It also means that a manuscript written in Newfoundland can be edited electronically in Boston and delivered electronically to Los Angeles.

The MAP (Manufacturing Automation Protocol) network developed for use at GM, Ford, and McDonnell Douglas allows different technical systems involved in manufacturing to be coordinated. The MAP users' organization, which in 1985 included 240 companies, recognized that incompatibility among systems had created a "communications monster." As a result, the organization has pressured vendors to develop computerized machinery that can be used as part of a system. The ideal is nonobstructed, fluid communication throughout the production process. At the GM Saginaw plant, it is estimated that switchovers for the production of axles for 1986 models using the MAP system will take ten minutes. In 1985, it took three days.

In a study called *Impact of Microcomputers on British Business* (1979), the National Computing Centre in Manchester, England, links effective network capability to the adoption of an "integrated workstation." Ultimately, says this study,

> the product which will do most to change the operation of the office will be an 'integrated workstation'—the entry and exit point for information on a digital, voice, data, text, and image network. In its fully developed form, this will replace the typewriter, the office copier, the telex machine, the computer terminal, and the desk calculator, and will provide the key terminal facility for electronic

mail. . . . It will also have access to public data banks and information systems.[3]

Not only will the integrated workstation streamline work flow, but it will enable data to be transmitted to other sites. Limitations of time and distance will no longer impede work, and computerized cottage industries will have the same access to data as multinational corporations. Through the National Library of Medicine's IAIMS Program (Integrated Academic Information Management Systems), it is already possible to see the vast improvements this kind of system makes in information availability and access. Through the IAIMS Program, medical service delivery and research are expedited by information programmed to address specific needs. For example, at the University of Utah Medical Center it is possible to cross-reference medical information with a large genealogical database and conduct epidemiological studies that show the relationship between the country of origin of immigrants to the United States and the inheritance of diseases.

The Dark Side

By their nature and formidable capability, computers are changing the workplace. A great deal of labor is now organized to work with the computer rather than to direct it. As a result, workers are not only losing control over their work, they are also losing their understanding of it. Many high-tech workers today do not know why certain designs are considered more effective than others. As computers control more aspects of work, the gap between knowledge workers, who control the technology, and line workers, who are its custodians, will become wider. In the process, a number of skilled jobs will simply disappear, as intelligent machines become capable of performing more tasks with a competence that exceeds that of human workers. Already, draftsmen have been replaced by CAD (computer-assisted design) systems, telephone technicians by multilink transmission systems, and production line welders by robots. Recently, the tomato growers of California started using a robot called the Tomato Harvester to pick tomatoes. The implications for the migrant workforce are devastating: The number of farm jobs in the tomato industry in California dropped from more than 40,000

in the 1970s to 8,000 in the early 1980s. There are also negative implications for those who like the taste and texture of tomatoes, since growers are now concentrating on breeding plants tough enough to withstand the robot picker.

The plight of migrant workers raises an interesting question about the costs and benefits of automation. How are tasks targeted for computer application? The economics of adopting high technology, of course, depend on certain production-specific factors. How much risk is involved in a specific task? Robots clearly are superior to humans for handling toxic wastes. One designer pioneered the use of robots to move inventories in deep-freeze warehouses. Immune to extreme cold, the robots were subject to considerably less danger and discomfort than human inventory clerks in the same situation. Work situations involving hazardous conditions are, therefore, one area where high-technology substitutions unquestionably improve the overall safety of the workplace. Savings in the costs of extra pay for hazardous and unpleasant conditions, as well as of workplace insurance and disability, are other obvious benefits in this sector.

Another production-specific factor is speed and accuracy. The utilization of computers and robots in industry was initially geared toward simple repetitive tasks that could be systematized at high speed. Spot welding currently accounts for the use of thirty-five to forty-five percent of all the robots in the United States. At General Motors, which has been one of the strongest advocates of robotics, each first-generation industrial robot purchased in the early 1980s eliminated an average of 1.7 to 2.7 jobs in plants that operate on a three-shift, twenty-four-hour-a-day schedule. Taking into account a British study which suggests that for every 2.5 jobs lost to robotics only four-fifths of a new job is created, it is easy to see some of the economic benefits of the new technology for employers. Researchers have estimated that by 1990 the United States will have lost 100,000 to 200,000 manufacturing jobs to robotics.[4]

But automation does not necessarily produce immediate economic benefits. In work situations where precision and accuracy are the basis of cost-profit ratios, computerized equipment can increase profits—but only if it is introduced with adequate knowledge and training for those who maintain and use it. Several recent anecdotal accounts indicate that without adequate knowledge of the new technology, workers responsible for it become unintentional saboteurs.

Companies have consistently overestimated the ease of transition from mechanized and computerized production, and have underestimated the skills and training the new technology demands, with costly results in terms of time and wasted materials. A well-documented example is the flexible manufacturing system at Caterpillar Tractor. According to Jack Hollingum, writing in *The Engineer*, it took four years from the time Caterpillar ordered the system in 1971 for it to become operational.[5]

Furthermore, the capital investment that new technology requires (estimated at $51 billion for computerized equipment in 1984) can lock corporations and institutions into rigid, inefficient timetables. Too often, the complexity involved in designing an automated system is underestimated. Close examination of even the most basic tasks reveals that the actual work process is much more complicated than the sum of its parts. This has been one of the strongest challenges in the transition to an automated work environment.

Another concern is the degree to which workers whose tasks are being automated are able or willing to provide information for designers and programmers. The hidden aspects of job competence are consistently the most difficult to teach a machine. These are often highly variable and unquantifiable: they involve experience, trial and error, and sense, touch, or knack. A case in point was the Campbell Soup Company's decision to create an artificial intelligence program for the master soup-taster's job. While this might seem to be a straightforward set of tasks, it took the soup-taster and a programmer more than three months to map out the job so that it could be programmed.

What is true of workers is also true of management. Recent studies have indicated that a good deal of what managers do is gather information informally. Exchanges in the corridor are common, and power is brokered in ways that resist documentation. This makes it unlikely that electronic management systems will ever be designed with all the capabilities of human managers.

Designers and programmers also often encounter resistance when they are researching system specifications. Workers, including managerial workers, are suspicious of computerization, especially when it might threaten their job security, minimize their skills, or challenge their judgment. Misrepresentation of work organization or data and outright refusal to adopt the technology are all aspects of such re-

sistance. A familiar example is the middle-level manager who resists using a word-processing pool in order to retain a secretary. Not only are two hierarchically different positions involved, but also an intricate web of status, dependency, and habit.

Toward the Postindustrial Society

Overall, the benefits to the company of automation outweigh the costs, particularly when planning takes into account indirect costs, such as maintenance and upgrading. The reticence American industry has shown recently toward the robotics market is an example of the pendulum swing between overenthusiasm and its natural fallout, disillusionment. The euphoria of the computer age has begun to give way to realism regarding what the technology can do. However, a new thrust is expected, built on such sophisticated technologies as the MAP network, once such systems demonstrate their true capabilities.

Industrial competition from Japan and Western Europe will also continue to fuel U.S. research and development. Japan is now ahead of the United States in the successful use of automation in certain industries, and this is weakening our international competitive position. Given the economics of the marketplace, according to *Forbes* and *Fortune* and many of their readers, strategic implementation of high technology is critical. American manufacturing industries recognize the threat and are acting. In part, the result is job displacement. In a provocative editorial on why the United States won't become a permanent debtor nation, *Forbes* editor Howard Banks writes: "Robots and computers are achieving efficiencies that reduce the importance of wage costs in calculating where to locate a plant."[6] His thesis is that, with high technology, industry will not have to continue to relocate to cheaper labor markets in the third world in order to compete in the international market.

Most businesses have a straightforward motive for automating: rationalizing production and reducing the number of people required to meet production goals. The first allows centralization of control, a process complementary to the bipolarization of the workplace. The second allows cost reductions in wages and benefits, and raises the thorny social question of who will assume responsibility for the unemployed and underemployed sacrificed in the process.

2

Postindustrial Society Reconsidered

T HE decades of the 1950s and 1960s were rocked by events that resulted in accelerating social change. Suburbanization changed the landscape and reoriented the American family. Television, which caught on rapidly in the 1950s, heralded the era of the "mass society." The shock of Sputnik and a sharpened sense of competition with the Russians raised concern about educational reform and support for military technology. The civil rights movement, and the U.S. Supreme Court's landmark decision in *Brown v. Board of Education*, brought Americans to grips with segregation and institutionalized racism. With the publication of Michael Harrington's *The Other Americans* in 1962, society was reminded of yet another sizable group of outsiders—the structurally unemployed.

On reflection, it seems that what Americans thought and felt during the twenty years that followed World War II alternated between two poles. At one extreme was the conviction that anything was possible: science and technology could be applied to any situation to create Utopia. Work would be automated. Society would devise some means of managing the psychological problems created when work was no longer necessary and could therefore not provide the basis for identity. The social world was the new frontier, and it would be conquered with the same sense of manifest destiny as the geographical frontier had been a century before. In fact, the programs of the War on Poverty were united under the rubric "the new frontier."

At the other extreme—expressed most vividly in fiction—was the concern that this utopian future would somehow turn out to be

a nightmare. Rather than society creating a framework for the satisfying use of leisure, individuals would become alienated and their lives would become barren. Those who looked ahead with a jaundiced eye doubted that everyone would benefit from technology. They were concerned about who would be given opportunities to work, how status would be determined, and who would exercise power. They were not convinced that technology would do much to alter human nature, and doubted that the rewards of the new age would be shared any more equitably than the rewards of the old.

The productivity potential of computers and high-technology robotics created concern among realists and alarm among futurists, who foresaw a massive increase in time spent away from the workplace. The ideal of a job for everyone gave way to schemes for rationing what would be left. Education became the primary entrance requirement, with great emphasis on higher education. Buttressed by the G.I. Bill after World War II, which enabled 2.3 million men to attend college, a college education became first a possibility and then a necessary badge for the upwardly mobile children of working parents.[1]

Amid changing concepts of education, work, and leisure, America was emerging as a consumer society. The new suburban homes were filled with labor-saving appliances. There were—or so we are led to believe—two cars in every garage and the suburban housewife experienced less drudgery and had more time to think. The marketing of the American Dream occurred across all social classes. Some sociologists linked the consumerism of the emerging blue-collar middle class with their desire to distance themselves from the remembered constraints of poverty. A sharp division emerged between the poor and the rest of society, and an "underclass" emerged, whose members, trapped in the "culture of poverty," enjoyed none of the new social mobility and participated only minimally in the new consumerism.

Poverty was frighteningly real to this generation of workers, most of whom had survived the hard times of the Depression. The affluence that fueled the new consumerism, however, reflected the substantial gains won by these working-class Americans during World War II. But affluence was not universal, and higher wage scales increased the distance between most of the workforce and the unemployed, underemployed, and marginally employed.[2]

Although a segment of the population was excluded from the society of affluence, consumerism was considered not only normal but a sign of membership in the culture. It was a realization of the American Dream, which embodied a rigid formula for success—hard work, education, ambition, and a need for recognition and achievement. The importance of work as the mediator of identity had its genesis in the industrial and technological revolution.

But the newly affluent workers of the 1950s found few rewards in the industrialized workplace other than material ones. As social critic David Reisman wrote in 1958: "The great victory of modern industry is that even its victims, the bored workers, cannot imagine any other way of organizing work now that the tradition of the early nineteenth-century Luddites, who smashed machines, has disappeared with the general acceptance of progress."[3] But in study after study, workers of the 1950s did not appear all that enchanted. A survey conducted by Morse and Weiss in the mid-1950s showed that some eighty percent of the industrial workers interviewed stated that they would keep working for lack of alternatives, but took little satisfaction from it. Descriptions of blue-collar work in the 1950s included pictures of drudgery, danger, and uncertainty. Attempts to make work more meaningful often involved making it more time-consuming and gregarious, but never questioned its structure.

By contrast, their bosses, the engineers and managers, reported high levels of job satisfaction.[4] The organization man was so totally integrated into the corporate culture that his work life and his personal life were one. This was the stereotype Sloan Wilson portrayed in *The Man in the Gray Flannel Suit*, a best-selling novel of the period.

The Eclipse of Industrialism

Management experts generally ignored worker attitudes and focused on the potential of substituting labor-saving and transforming technologies for bored or discontented workers. Technology provided a vision of the absolute mechanization and rationalization of production. Within the information and service sectors, the organization of labor and the goals of production were not so clear and the benefits were more difficult to specify. During this period of reassessment, it became clear that a society based on industrialism was changing and that the institutions which had evolved to support it were no

longer adequate. Small firms managed by independent entrepreneurs were giving way to giant corporations and large-scale bureaucracy.

What was underway was the eclipse of industrialism. Accelerated technological change was a vital part of the shift not only to a new basis of production but to a new social order. Slowly, the idealized future model emerged as a system in which the management of information was to be the key to corporate success. The postindustrial society was to be one that put power in the hands of the information handlers.

Industrialism and postindustrialism do not only take different approaches. They have different goals. In the workplace, a postindustrial society uses high technology and sophisticated communications to *replace*, not to *assist*, human labor. In society, it seeks to impose order. Methods developed for science and technology are transferred to other sectors because a postindustrial society, in order to function as the technologists theorize, requires central planning and absolute control.

Dozens of government and private commissions met and developed plans to explore what the institutional framework of the new society might be. The same problems we see today—displacement, the meaning of work, and the need for corporate programs responsive to radically changing economic and social conditions—were at the heart of these discussions. And the solutions proposed then are still mostly valid. But, as history has shown, they will not occur spontaneously.

The Idea of the Postindustrial Society

The idea of a postindustrial society was clearly predicated on the assumption that the changes taking place in the economy, politics, and culture were radical and new. A complex evolutionary model of this unprecedented era would involve new patterns and styles in health, education, government, and business. A vaguely utopian concept of the future included the advent of a leisure-oriented society and held the promise of escape from poverty and misery. It was formulated as a benevolent alternative to more sinister views of the alliance of science and the military. Interestingly, the term "postindustrial" was coined at roughly the same time that President Eisen-

hower warned about the dangers of an overly powerful military-industrial complex.

In the late 1950s, when it first came into use, "postindustrial" was used interchangeably with such terms as "information society," "cybernation," "postcapitalist society," and "postmodern."[5] It conveyed a sense that some new set of social arrangements was in the making, with parameters the forecasters could not quite pin down. "Post" conveyed a historical difference in time. "Industrial" conjured up the large-scale factories and plants of the nineteenth and early twentieth centuries. Current usage retains the sense of society in transition, but locates science, knowledge, and technical research at the center.

One of the most influential treatments of the subject was Daniel Bell's *The Coming of Post-Industrial Society* (1973). Bell examined five dimensions of this new society:

1. The changes in the economic sector as we shift from a goods-producing to a service-producing economy.

2. The preeminence of technical and professional occupations. Bell heralds the rise of power of technocrats who use technology to broker power, and technicians who apply it to change work allocation.

3. The ways in which knowledge and values are applied to the direction of social institutions: The postindustrial era will utilize game theory and similar theoretical approaches to make practical decisions.

4. The goal-setting process for the planning and control of technological growth: how methods of technology assessment are designed, and the creation of political mechanisms to authorize assessment and define criteria for regulation and intervention.

5. A new decision structure utilizing "intellectual technology" or the "substitution of algorithms (problem-solving rules) for intuitive judgments." Such rules would succeed in making societal decisions "rational."[6]

But whether it is called postindustrial or postmodern, the last half of the twentieth century is perceived as a time of radical change. The nineteenth century witnessed the advent of widespread mech-

anization. The first half of this century was a laboratory for efficiency: machinery and organized production allowed optimal use of time and energy. Refinements on the assembly line proved that time was money, as a constant number of workers produced more and more goods. All these approaches furthered the general goal of increased production for a seemingly limitless market.

In the postindustrial society, this is still the technocrats' dream. The basic difference now is that the market is finite and the competition universal. In such a world, the strategy changes from one dealing with constant or increasing numbers of workers producing more goods to one in which fewer and fewer workers produce goods that can be competitively priced.

Efficiency is a compelling argument. It conforms to the work ethic and to the notion that time is money.[7] It appeals strongly to the profit motive. In the United States, efficiency produced a society of abundance in which speculation about alternative reward systems could flourish. But doing business in such a world required the application of technical knowledge to large-scale problem solving. In the two decades following World War II, when the United States dominated the world economy, increasing importance was placed on rationality. In this context, the computer becomes the ideal tool, since it is capable of manipulating information with no concern other than the rationality of its output.

Leontief's Input-Output Model

In a 1983 speech at a National Academy of Engineering symposium, Wassily Leontief assessed the problems and opportunities of automation: "Any worker who now performs his task by following specific instructions can, in principle, be replaced by a machine. That means that the role of humans as the most important factor of production is bound to diminish—in the same way that the role of horses in agricultural production was first diminished and then eliminated by the introduction of tractors."[8]

Leontief's comparison between the horse and the worker illustrates the ways in which machines can replace animal energy. Significantly, the tractor did not replace just horses. Agriculture employed almost sixty percent of the workforce in 1900, today it accounts for less than four percent, and the agricultural policies of

the last few years promise further reduction. The devastation created by farm foreclosures is a concrete example of how policy translates into adversity. According to Andrew H. Malcolm on February 4, 1987, in *The New York Times*, "Around the clock, through the year, 180 times a day now, another American farm disappears, another victim of consolidation, changing economics, poor management, bad luck, high interest rates, low crop prices, emotional and financial despair, or some of each."[9] Simultaneously, farm productivity has soared, illustrating the potential of technology to transform both work and production. What Leontief's input-output approach provides is a description of the flow of goods and services and the underlying economic structure. It takes into account direct and indirect effects of changes in labor, capital, and infrastructure, and the interactions among these.

In the spring of 1985, the Congressional Research Service (CRS) released a study called "The Impacts of Automation on Employment, 1963–2000," by Leontief and Faye Duchin.[10] Based on estimates for eighty-nine individual industries and fifty-three different occupations, it maps the directions the labor force will follow:

1. The impact of the microelectronics revolution in the United States will most likely not be as earth-shattering as many futurists have argued.

2. When compared to the changes wrought in Europe by the Industrial Revolution, the changing division of labor will be like the changes that took place in Europe before 1820. Barring dramatic developments in artificial intelligence or cybernetics, for example, it is possible that more jobs will be created than are destroyed.[11]

3. Regarding industry-specific shifts, the evidence indicates that certain groups, such as female clerical workers, could be hard hit. Clerical workers, who represented 17.8 percent of the labor force in 1978, may drop to 11.4 percent by the year 2000. Managers could also suffer some decline. Sales workers are expected to remain at about the same levels as in 1978. The proportion of workers employed in service industries is expected to increase. Estimates are that the number of service industry employees could account for almost 15 percent of all jobs by 2000.[12]

4. While the data indicate that there could be enough jobs, there are questions about the *quality* of those jobs. The Leontief-Duchin data lend support to those concerned with the potential mismatch between skills and workers.[13]

The Leisure Society

The concept of leisure as it has developed in the postindustrial debate is also a focus of elaborate planning and rationality. Concern about how people will spend their idle hours has been a major theme of American thought since Puritan times. The division of the day into work and nonwork time has always tempted social engineers, who hold that the discipline of the workplace will not necessarily remain intact unless it is reinforced by some program of organized and supervised leisure.

Victorian society, both in Europe and the United States, was preoccupied by the need to develop middle-class habits in the working classes. The urban parks movement of the nineteenth century, which resulted in New York's Central Park and the many local playgrounds in American cities, was promoted by civic leaders concerned with urban order. Left to their own devices, it was assumed, the laboring classes might become the dangerous classes.[14] In Victorian societies, where industry and education were the keys to social mobility, leisure time was the only aspect of daily life left to be organized. This was not just crowd control: it was linked to the genuine belief that through literacy and instruction in the responsibilities of citizenship, it was possible to produce a mature and disciplined working class. The middle classes could strike a proper balance for themselves, with their cultural societies and commitment to betterment; the working classes needed a helping hand. Their beer gardens and dance halls were not conducive to self-improvement.

Mass entertainment changed much of this. By the 1950s, the missionary attitudes of settlement house social workers and local recreation councils had given way to the idea of leisure as free time and as a counterpoint to work time. The baby boom combined with suburban relocation to create a culture that sought luxury and self-fulfillment. The new technologies of the workplace were helping to free time for family, sports, hobbies—and television. The shift from the work ethic of the pre-World War II period to a leisure-consumer

society was diagnosed in 1958 by David Reisman: "For many people today, the sudden onrush of leisure is a version of technological unemployment: their education has not prepared them for it and the creation of new wants at their expense moves faster than their ability to order and assimilate these wants."[15] Leisure, formerly the province of the very rich (and the very poor), was becoming more of an option for all. It offered an antidote to the frustration and tedium of the workplace.

With all the concern for leisure and how to use it, one fact of everyday life stands out: Work and the time devoted to it is resilient. There has been little reduction in the length of the workday since the mid-1930s. Programs to reduce the number of days worked in order to maintain levels of employment have met with mixed reactions. As always, people need work as a source of definition and identity. Even during the 1970s, a decade characterized by preoccupation with the self and leisure experimentation, job-related identities were still all-important. A recent Roper Organization survey found that only one person in five places more emphasis on personal satisfaction and pleasure than on working hard and doing a good job. A 1983 American Council of Life Insurance survey found that eighty-five percent of those sampled believe that success in life depends on hard work.[16]

Evolution, Not Revolution

So rather than creating a new society with a new framework, we are simply evolving. The fine line between the creation of a new social order and refinement of the old one is hard to find, though tempting to announce. Leisure, like other attributes of the postindustrial society, is just one more aspect of that evolution. The old core values that make work central to identity are still very much with us. A 1972 U.S. Department of Health, Education and Welfare publication entitled *Work in America* put this well:

> [T]o be denied work is to be denied far more than the things that paid work buys; it is to be denied the ability to define and respect oneself.
>
> It is illusory to believe that if people were given sufficient funds most of them would stop working and become useless idlers.[17]

One example from the 1960s should be remembered today. In 1963 a contract negotiated by the United Steelworkers gave high-seniority workers a thirteen-week sabbatical every fifth year in lieu of a wage increase. While these thirteen weeks might have been used for self-improvement, for the most part they were not. The sabbatical was just a long vacation—workers returned to the same jobs with the same skills. But although the time did not necessarily lead to retraining, it did achieve one goal: the redistribution of work. It established a model for work sharing that could be instructive in today's environment of displacement and layoffs.

The resilience of work as a source of personal worth and income has led many critics to revise the idea of the leisure society. Since ours is becoming a service economy, the "business" of leisure is becoming an increasingly important employer. Health clubs, beauty care, and even antique stores and art galleries are booming. The list grows as our receptiveness to new and organized uses of free time becomes more developed. Work in the industries that serve the expanding upper-income groups—including today's affluent young professionals—has become a way of reinforcing class identity. Low-income service work versus high-income professional and managerial work comprises the additional dimension along which class lines are being drawn.

3

The Elusive Utopia: The Technological Forecasters

T HE futurists have focused relentlessly on the positive: The wonders are surely coming, the only question is how fast. When it comes to timetables for change, we have had a choice of several crystal balls. Bestsellers like Toffler's *The Third Wave* and Naisbitt's *Megatrends* have heavily influenced our attitudes about what will come next and provided a new vocabulary to describe it.

Although popular accounts have tended to stress the promise of technology, their predictions have rarely noted the very great distinction between what is technologically possible and what is politically, economically, or institutionally practical—or even desirable. These views of the technological future present a well-ordered, machine-dominated world where drudgery disappears and just about everyone lives happily ever after.

The writers assume that the social changes which will accompany the diffusion of technological innovations are inevitable and essentially positive. The distress of those who are left behind is perceived as something like society's pains. Toffler, for example, knits his analysis into a theory of revolution which argues that the seeds of the new society emerge from the exhaustion and erosion of the old.[1]

Public approval of technological innovations is an important step in getting them into the workplace and the home. A positive view of how life will be changed can help moderate resistance or ambivalence. Accounts that accentuate the promise of technology, espe-

cially those buttressed by statistical and trend analyses, carry the weight of authority and give legitimacy to the *idea* of constant change.

Of course, Toffler and the others acknowledge some costs and risks of technological change. However, they fail to consider too many uncertainties and unknowns—including how unprepared most of the current workforce is for the workplace of the future. There are also too many instances in which their predictions have already been dead wrong.

Utopian thought is certainly seductive: It offers security and the comfort that comes from the appearance of knowing where one is going. But beneath the surface lie troublesome questions. If half the workforce does twice the work in a steady or declining market, what will the other half do? Who will make the regulatory decisions that keep Utopia functioning? But these are problems barely considered in what the futurists propose.

Images of the Future: The Reign of the Computer

The impact of computerization and chip technology is central to futurist projections. All build from an analysis of the changing division of labor. Integral to their analysis is a look back to the thresholds of the Industrial Revolution—mechanical energy, the assembly line, the search for efficient ways of producing goods and services. According to a futurist scenario, computerization takes over the functions of the traditional worker, and networking makes possible adjustments in where and when people will work. Electronic miracles let us return to the piecework methods that characterized cottage industries in the preindustrial economy. The analogy here is to industries like weaving, which were a significant source of employment prior the the invention of mechanical looms.

At the macroeconomic level, these changes include the internationalization of economies. The nation-state, with its territorial and fiscal boundaries, no longer forms a viable economic entity. National economies are beginning to give way to international production systems. Research on corporations indicates an increasing tendency toward farming out work to cheaper labor markets. International trading patterns and currency markets also exert new pressures on political structures. Barter between corporations is one way in which jurisdictions are transcended. Here, for example, is one recent barter deal: magnesium from China is traded for coffee beans

from Brazil, which are in turn traded for handbags from Spain. The middleman is an American-based Fortune 500 corporation. Which government is in control?

Change also occurs within the social structure and mediates human relationships. Toffler, in *The Third Wave*, speaks about "the apparent incoherence of political life which is mirrored in personality disintegration."[2] The cult of the self, exploration with different types of therapies, and religious revivalism are all tied together. His third wave is the civilization that emerges as manufacturing gives way to service and information work when the computer organizes production.[3] There is a parallel theme: the cult of the self as a reminder that we are still flesh and blood, not just rational, cerebral beings.

Toffler assumes the electronic cottage, made possible by the computer, will be central to work, and he discusses marriage and family in terms of what will happen when husbands and wives are thrown together for longer periods of time. In his vision of the future, the shortage of full-time employment is not a big problem. It will become possible for husbands and wives to share the equivalent of one full-time job. Any number of employment arrangements are possible with home work stations. Toffler's emphasis, however, is on the repercussions of such new arrangements on the home and marital bond. More time together will demand new ways of relating. The family as we know it will also change as people choose from an array of options that include contract marriages, serial marriages, family clusters, commuter marriages, elderly cohabitation, and various kinship arrangements.[4]

High on the list of transitions the third wave brings is the growing importance of information as a commodity. Toffler makes a critical distinction between info-property and real property. "Info-property is not finite; I can use it and you can use it at the same time. In fact, the more people use it, the more information is likely to be generated."[5] Thus, such concepts as wealth and property will have to be reconsidered, as will the social relationships based on them.

Changes in work and the economy brought about by the shift from manufacturing to information are also the central themes of *Megatrends*, which lasted sixty weeks on *The New York Times* bestseller list.[6] Whereas Toffler presents the society of the future with a measured enthusiasm, Naisbitt's tone is totally upbeat: the key to success is to "go with it." While the future clearly represents a challenge,

"trends, like horses, are easier to ride in the direction they are already going."[7] The impact of high technology on the workplace can be seen in the ratio between high tech and high touch. High touch is the "counterbalancing human response" which ensures that the technology being introduced will not be rejected.[8]

Naisbitt's high touch metaphor links the emergence of the human potential movement, with its various therapies and self-improvement regimens, to the preeminence of high technology. The promise of the future is embodied in a modern version of the ancient Greek principle of balance. Naisbitt contends: "We must learn to balance the material wonders of technology with the spiritual demands of our human nature."[9]

Unlike Toffler, Naisbitt argues that the electronic cottage will not replace the office. "The utilization of electronic cottages will be very limited: People want to go to the office; people want to be with people."[10] In the large-scale workplace, high technology allows for accommodation to personal need. In his brief treatment of the unionized workforce, Naisbitt points out the asymmetry between historical demands of unions and the capabilities of computerized administration:

> Because we have the computer to keep track of it, an individual contributor to the pension plan can decide where that contribution is going to be invested. And that is one of the key reasons that unions are out of tune with the new computer-rich information society. The basic idea of a union is to ensure that everyone is treated the same. But now we all want to be treated differently.[11]

A picture of the future as compelling as Naisbitt's is tempting, but misleading. The gloomier side of deindustrialization and the fate of displaced workers should dampen our enthusiasm. Delays in the realization of a technological Utopia should also sharpen our skepticism, although delays may prove advantageous, since they allow time for self-correcting mechanisms to evolve.

Matching Naisbitt's enthusiasm for the coming era—but taking a think tank perspective—is Herman Kahn's *The Coming Boom*.[12] What distinguishes his analysis from Naisbitt's and Toffler's is that the data and argument are oriented not to a popular audience, but to policymakers. In a serious economic analysis of trends, Kahn presents a compelling argument for an optimistic view of the future.

According to him, the decades of the 1980s and the 1990s will have "relatively high and sustained" growth rates, due to the application of technologies and technological improvements. He also contends, correctly, that the lag between the invention of technology, its appearance on the market, and commercial application is diminishing.

Kahn would meet the challenges to labor and the world of work by reestablishing what he describes as an "ideology of progress."[13] Maintaining social coherence and a sense of mission are indispensable themes that run through *The Coming Boom* and most other futurist texts. They represent a belief that the next stage will be somehow better than the current one.

In Kahn's analysis, the growth rate of the labor force is one index that comes out negative.[14] On the positive side is the fact that the baby-boom generation has begun to mature. Kahn equates this with a tendency toward more conservative attitudes among a generation that had called into question most of the institutions of society, and a "nesting" lifestyle in which there is greater consumption of durable goods and housing. Together, these changes in values and consumption should help curb inflation. However, those able to consume according to their desires are only one part of the labor force.[15]

On the negative side are those unable to consume as they wish because there will be more new entrants to the labor market than new jobs. What about those who have been left behind by previous economic recoveries? Kahn and his colleagues at the Hudson Institute project a "secular trend toward higher unemployment." In their view, the chronically unemployed are less motivated than the boom-oriented job seeker to look for jobs or to accept the kinds of jobs that are available. Their estimate, which has proven low, is that acceptable unemployment will be between six and seven percent and that society will accommodate itself to this level.

With changes in employment patterns that result from voluntary part-time employment and discretionary entry and departure from the labor force, the terminology used to discuss labor force participation will need to change:

> The automatic classification of a person who leaves the labor force as a 'discouraged worker' is no longer accurate. In fact, the classification probably applies to less than half the people who move

out of the labor force. The term, which came out of the depression and raises the image of an individual who was desperately looking for a job, is not representative of today's teenager, student, wife or part-time worker who goes into or out of the work force for personal reasons.[16]

Kahn draws important conclusions when he questions the types of unemployment categories and statistics that we use to build analyses of the labor force. Voluntary leaves and part-time employment are options for some and will continue to allow flexibility for those who can afford them. However, to assume that all who are working part-time or who are not working for a period of time have reached that state voluntarily is simplistically wrongheaded. Reading Kahn, the contradictions of a conservative ideology become apparent. If everyone who wants to succeed will succeed, how do we account for the fact that half the unemployed arrive at that state involuntarily? One unemployed auto worker put it this way: "I did everything right, worked by the rules, and still lost my job."

In a recent piece, "Reaganomics in Retrospect," economist Isabel V. Sawhill brings into focus the political valence of Kahn's concept of voluntary unemployment:

> With respect to unemployment, conservatives argue that most of it is voluntary. The problem comes in defining the word "voluntary." If by voluntary we mean that people could find work *on some terms* (for instance, work well below their customary wage, in a different location, or in a new line of work), then it is probably true that much unemployment is voluntary. So the debate about the seriousness of unemployment, from a social welfare perspective, revolves around beliefs about what kinds of compromises workers should be expected to make in their search for work.[17]

The argument implicit in the conservative definition of voluntary unemployment is that expectations regarding quality of work are secondary to earning a wage.

Regarding changes and technological effects on labor, Kahn argues for careful attention to entrepreneurship, productivity, capital accumulation, and investment. Quoting GM executive Roger Smith, he highlights the promise of high technology: "Every time the cost of labor goes up $1 an hour, one thousand more robots become

economical."[18] Kahn predicts a rate of increase in the use of robots for factory work of about thirty-five percent a year in the 1980s. If this holds, GM's labor force will consume far less coffee and much more electricity by the year 1990. He also sees widespread use of sophisticated office technologies with extensive networking capabilities and a shift from paper to electronic files. Widespread use of optical scanners and devices that operate by natural voice command will complete the changeover.[19]

The direction for technology will involve the formulation *C*-4 *I*-2, in which *C* stands for command, control, communications, and computing, and *I* stands for information and intelligence. The distinctions between information and intelligence are instructive. Information refers to routine data, whereas intelligence is "knowledge about events and people that may be conjectural, intuitive, private and/or obtained at random, unofficially or clandestinely." Intelligence, the more volatile element of the formulation, is responsible for generating change.[20]

Kahn's arguments are tempting in that they make strategic use of the American value system, with its emphasis on the prerogative of the individual and the belief that the future will offer something better. Unlike Toffler and Naisbitt, his view is as much evolution as revolution. The America of the frontier and the heyday of entrepreneurial capitalism are necessary antecedents for the coming boom. They represent benchmarks in our national resilience and capacity for development.

Collectively and individually, these three prophets see imminent and all-encompassing change resulting from the computer and its spinoffs. Economist Robert J. Samuelson in a *Newsweek* article of September 9, 1985, called "Our Computerized Society," adds a note of needed realism:

> Computers are only now emerging from their mid-ditch era. Like autos in the 1920's their growth has been chaotic; like autos, the questions of social impact are unanswered. It remains an open issue of how wisely we exploit computers' potential. Just because a technology can do something doesn't mean it's socially or economically useful. But our predictive powers are modest.[21]

Samuelson's point is well taken: Our sense of what computers' long-range impact will be is still fuzzy.

Some Hard Questions

The futurists all foresee improvements in productivity and an economy reconfigured to exchange services and information. They acknowledge as inevitable the tensions between the old industrial order with centralized authority and the smaller-scale organizations that require professional involvement and team spirit. In a dangerous leap of faith, however, they all accept that work will become relatively less important to most people.

A. R. Martin, a chairman of the American Psychiatric Association Committee on Leisure Time and Its Use, in the late 1950s put forward a different view that current studies reinforce:

> We must face the fact that a great majority of our people are not emotionally and psychologically ready for free time. This results in unhealthy adaptation which finds expression in a wide range of sociopathologic and psychopathologic states. Among the social symptoms of this maladaption to free time are: low morale, civilian unrest, subversiveness and rebellion.[22]

With robotics and computerized office management systems, the need for human labor will inevitably decrease. The 900 percent increases in productivity reported in 1984 among draftsmen using one of New York City's computer-assisted design (CAD) systems is one example.[23] Since even conservative estimates project a ten percent across-the-board reduction in jobs, we must either condition people not to value work so highly or find more creative ways of dividing up what work is left.

One means of addressing this issue is to assess which parts of society will have to change to accommodate new work patterns. An example can be found in France. In 1977, the French government supported a far-reaching study of the impact of computerization on society. The Nora and Minc report concluded: "The general trend of society will require a decreasing amount of productive work. In terms of total volume, this evolution is inevitable."[24] Like the American futurists (with the exception of Kahn), the French authors put the challenge of stabilizing the new information economies onto the political and cultural institutions of society.

They argue that social instability is one potential consequence of diminishing the significance of work, because it reduces both the

time work consumes and the value it bestows. Diminishing the role of work will serve to intensify conflicts within society and create the possibility of disruptive activities. Their contention is that the changes taking place now will give impetus to social movements like the student movement of 1968 and the more current neighborhood movements.

The student strikes of late 1986 in France and Spain offer concrete evidence of the logic of this argument. It is significant that among the issues raised by the students who took to the streets in Paris in November 1986 was concern over the value of their education and whether they would find employment once they had graduated from the university.[25]

In general, Europeans have been more willing to identify and discuss such negative aspects of automation as displacement and lack of training. For them, the threat to the blue-collar middle class has led to a more interventionist approach. French, German, and Scandinavian trade unions, much stronger than their U.S. counterparts, have worked closely with governments to ensure job protection through time-sharing and early retirement programs, as well as through fairly substantial budgets for retraining programs (four percent of Sweden's national budget is devoted to employment training).

One cultural effect of increased computer use, noted by Nora and Minc, is on language: "The enormous spread of data processing that will occur where access vocabulary is standardized will affect language and syntax."[26] They argue that once computerized languages become the mainstay of communication, they will emphasize and legitimize already existing patterns of social inequality.[27]

For example, in the United States, the kids most likely to get home computers are those whose parents can afford them and believe the advertising that equates failure to provide home computers for children with their subsequent failure in life. The Nora and Minc report places the responsibility where it belongs. Children with varying levels of opportunity to learn the new technology will all have to live in the new culture. To the extent that unequal opportunity to learn results in occupational and cultural handicaps, public education must be the equalizer. Failure to address this issue will result in predictable failure in the job market for substantial numbers of children from low-income families. Then the employment problem will become a political problem.

One futurist writer makes a strong case for basing any debate on the future on humanistic concerns. Clark Kerr ends his book *The Future of Industrial Societies* with a challenge from Bertrand Russell that was written in 1923: "The important fact of the present time is not the struggle between capitalism and socialism but the struggle between industrial civilization and humanity."[28] Kerr calls for new visions of the future that include "adjustments between efficiency and equality, between individual freedom and economic stability." He calls attention to the overenthusiasm with which plans for the future were developed in the 1960s and the pessimism which seems to be developing in the 1980s. By raising challenges at the level of humanity, he achieves the objective many futurists lose sight of in their excitement about technology. The economy may be abstract, but its participants exist in a real world filled with real consequences.

4

The Corporation, Management, and Technology

C ORPORATE America, as we have noted, is awash in computers and high tech. To see what this means in practice, it might be useful to look at three scenarios. The first concerns the Ford Motor Company, pioneer of the assembly line and a bastion of corporate America, and how it plans to overcome competition from overseas. A 1985 *Forbes* article, "We're Going To Do Just Fine," which considered how Ford might confront both international competition and high production costs in the United States, put reducing payrolls first among the strategies likely to enhance corporate longevity. Following closely were raising productivity, squeezing suppliers, and exporting production. To these, Ford's CEO Don Peterson adds consumer loyalty and a reputation for quality.

Thus far, the company has done the job where it counts most: between 1979 and 1985 Ford cut its workforce from 239,000 to 170,000. By 1990, another twenty to twenty-five percent of its hourly and salaried workers is expected to go.[1]

Scenario two concerns the trimming of America's corporate middle management. E. Jennings, professor of management at Michigan State University, estimates that eighty-nine of the one hundred largest U.S. corporations are involved in total corporate reorganization. Popularly termed "downsizing," this is simply an across-the-board effort to reduce staff size through a variety of means. Reductions

often focus most sharply on the organizations' middle ranks, as executives seek to have their organizations perceived as "lean and mean."

There are many approaches to downsizing. Dupont, which had hoped to eliminate 15,000 positions over three years using early retirement, was able to reduce its managerial force by 11,200 in 1985 alone, the first year of the program. However, encouraging early retirement through enhanced benefits is not only costly, but it can leave corporations with too many of the managers they would like to lose and too few of those they would like to keep. Since anyone eligible may opt to leave, whole departments can be depleted, and the most valuable employees may be the first to go.

A less-benign means of downsizing is to "stress out" targeted personnel through competitiveness, inordinate demands, and contradictory directives. Anthony Carnevale of the American Society of Training and Development finds that such methods may prove counterproductive. "The people we expect to be the most creative and flexible now work in the most insecure environment, and I don't think that's a good mix. Insecurity reduces productivity since it makes people more political, more afraid to take chances." Clearly, this approach created substantial morale problems as well.[2]

The effects of downsizing are evident in Bureau of Labor Statistics aggregate data. A recent study estimates that between 1981 and 1986 almost 500,000 "executive, administrative and managerial workers" lost jobs that they had held for at least three years. Compared with 64 percent of displaced blue-collar and factory workers, 72 percent of these upper-tier workers found new jobs. Although the rate of unemployment for executive workers in 1986 was 2.6 percent as compared to an overall unemployment rate of 7 percent, this provided little solace for the 28 percent of displaced managers who had not found new jobs. They were expelled from a world that they formerly helped shape.[3]

Scenario three looks ahead to a growing mismatch between skills and demand in employable populations. A recent International Labor Organization report concluded that, while the immediate impact of high technology on the labor force may be small, the long-term effects will drastically change job distribution. The result will be "a mismatch between the supply of and demand for skills on the labor market. This sectoral distribution of employment may also affect the distribution of employment between countries."[4]

These scenarios reflect dramatic changes in corporate staff patterns, reevaluation of the pyramid management structure, and reallocation of jobs by sector. Clearly, they run against the grain of an American corporate mythology captured in the 1960s classic *The Organization Man*. But this book is now considered a monument to corporate obsolescence, at least by the *In Search of Excellence* generation of managers, who have learned to distrust the corporate hierarchy. To these men and women, the executive of *The Organization Man* was not only too respectful of the corporate structure, he was also too docile within it.

The rhetoric that profiles the kind of managers who can give corporations access to excellence includes terms like "maverick," "experimenter," and "decisive risk-taker." The frontier mentality has been rediscovered by American corporate culture in its scramble to refit itself to ensure survival in the computer age.

The Impact of Computer Technology: Decentralization and Downsizing

The impact of computer technology on the organizational structure of corporate America is difficult to separate from the ideology and rhetoric used to describe it. One unambiguous measure, however, is found in staff patterns.

Computerization as a means of reorganizing the division of labor is most evident in the manufacturing, design, and information sectors. The impact of numerical control machines on the production of machine-cut parts has been documented by former MIT labor analyst Harley Shaiken.[5] Not only have craftsmen, like tool and die makers, been reduced to overseers of machines, but the working relationships among engineers, designers, and machinists have also changed. Their former collaboration has been replaced by a fixed division of responsibility for different parts of the production process. This separation was highlighted in the 1985 strike against Pratt and Whitney in the New Haven area. Striking machinists had no irreconcilable economic complaints against the aircraft engine producer. As one picketer put it: "It's not that we are not paid well, it's that they don't recognize the quality and responsibility of our work."

Parallel examples are found in the production patterns of high-technology industries like those in Silicon Valley. There, too, tasks

requiring the greatest skill have been passed from workers to machines programmed to perform with an accuracy and precision that human workers cannot hope to achieve.

In certain heavy industries, however, the adoption of high technology and robots demands high degrees of coordination in the production process. There, an opposite pattern has emerged. The automobile industry, for example, is moving away from assembly line organization, where each worker performs only one simple task. The alternative is a team or cell model, in which more worker discretion is exercised and a process is followed from start to finish. In the case of the General Motors Saturn plant, this is being accompanied by a reduction in the number of job classifications from more than one hundred to between four and six.

The team or work group approach borrows heavily from Japanese production processes, which attempt to foster worker interest in quality control and to encourage the assumption of responsibility for the finished product. This approach also often includes some degree of worker participation in planning and goal setting. But whether high technology leads to less worker responsibility and involvement or more, the intent and the effect is the same—more production with fewer workers.

In design, where the computer has to a large extent taken over the task of executing drawings formerly done by draftsmen, the result has been to eliminate costly delays, as well as costly employees. Computer-assisted design technologies not only expedite the design process but, when combined with effective networking and telecommunications capability, also permit design changes to be executed almost immediately.

In the information industries, the most visible impact of computer technology has been in the office, where the computer terminal is replacing the typewriter, the mailbox, the messenger, the file clerk, and a good many secretaries as well. Work is rationalized so that every function can be accomplished through the computing system. The result is the elimination of jobs as well as whole categories of office workers. Ease of access and immediate electronic transmission also reduce middle-rank information-processing positions. With computerized data analysis programs, many of the functions once assigned to middle managers can now be accomplished by executives themselves with desk-top computers.

The Monitoring Issue: Demoralization

Even a superficial view of the corporate structure reveals that re-organization is taking place from top to bottom. Change in the division of labor on the production floor is but one level of corporate reconfiguration. The other is found in management, where the familiar pyramid structure is giving way to flatter organizations that substitute entrepreneurship and team approaches for the top-down hierarchy. Not all of these approaches are compatible, but they share a common emphasis on decentralization.

From the perspective of a futurist like John Naisbitt, new organizational arrangements are the key to "reinventing" the corporation.[6] Initially put forward in the 1950s by such theorists as Likert and McGregor, whom Naisbitt refers to as the "new corporate humanists," these theories did not achieve their present prominence until the 1970s, when new economic conditions, including the oil crisis and the declining American industrial base, created the atmosphere for change.

The 1980s saw an even bigger threat: more intense and pervasive international competition. By reinventing the corporation, American business is, according to Naisbitt, reasserting its position in the world economic market.[7]

Despite such innovations as quality work circles and participative management, however, power is still unequally distributed. The myth of the corporate family has always barely masked the dynamics of competition and conflict that characterize environments where rewards go to the few. In the new corporate environment, rewards go to even fewer.

Office automation effectively displaces the resolution of conflict from the human arena to the technological one. Changes often reflect the needs of technology, not the needs of people. Furthermore, the computer, with its ready command of information and its comprehensive tracking systems, presents the possibility of subjecting all members of the corporation to close control.

Robert Howard, in *Brave New Workplace*, sees computer tracking of work activity as a refinement of Taylor's time-motion studies in the early part of the century.[8] The computer replaces the efficiency expert or personnel evaluator in gauging the complex relationship between input and output that defines productivity.

While these calculations have always been straightforward for lower-level positions in the corporate world, the ambiguity involved in managerial and executive work has made productivity in these positions almost immune to quantification. With the development of management information systems, management resource planning, and different computer-related measurements of productivity, higher-level positions become subject to machine-trackable accountability. Thus, a new genre of corporate discipline is developing, as is a more rationalized procedure of setting goals.

Studies now regularly document the stress felt by managers as more of their work is put on-line. The freedoms of higher-level work, which include self-monitoring and a certain degree of autonomy, are being eroded by automation. Research by psychologist Shoshana Zuboff shows that managerial workers find the electronic manipulation of symbols—working on inventories, for example, without any sense of what the symbols represent—a frustrating and demoralizing experience. Furthermore, loss of control over how work is done leads to stress. According to Michael J. Smith, chief of motivation and stress research at the National Institute of Occupational Safety and Health: "Monitoring is the sleeper issue of the next decade."[9]

Measured in terms of human stress, frustration, and demoralization, the costs of automating corporate America are pretty high. Stanley Hyman, a Washington area personnel consultant, translates the shift in work, with its accompanying contraction of middle management, into a "violation of implicit lifetime contracts."[10] Depending on when managers entered the corporate world, they have certain expectations about the type and quality of their work, and about career patterns. By changing the rules, the new corporate structures are destroying the former stability of middle management. The question of maintaining a stable corporate environment while trimming fat has always plagued management consultants; now it is becoming a concern in boardrooms.

The Computer as Regulator: The New Rules of the Game

In information-related jobs, the linking of different operations through a technological network not only expedites work, but permits more

effective and efficient machine-to-machine communication. Jokes about the "Tower of Babel" of information systems are becoming dated as new international communications technologies are improved and successfully linked to local area networks (LANs).

In this environment, such manifestations of corporate decentralization as farming work out to cottage industry operations becomes possible. Blue Cross, for example, processes massive batches of claims through telephone modem linkups from piece workers operating from home work stations. Supervision and management are conducted by review of machine-based records.[11] Indeed, ease of access to elaborate electronic files and periodic program checks are shifting the emphasis of supervision away from personal evaluation to machine-based review.

This new style of review is effective both at corporate facilities and off-site, and it is being adopted extensively at the same time that most corporations are experimenting with more humanistic forms of management. One of the hurdles to more effective "transparent review," as it is called, is the threat of inhibiting the kind of participation and alliance of interests that new management theories recommend. Attempting to integrate workers further into the corporate world, by tying their interests to that of the corporation, and simultaneously subjecting them to depersonalizing review procedures, clearly breed mistrust and encourage debate on the status and meaning of work.

Employees, already suspicious of management's real agenda, are quick to find ways to circumvent review procedures. Tinkering with the system by reprogramming to change the indicators of performance is just one example. Another is a Parkinson's law phenomenon—the generation of work to fill the available time. In the electronic office, for example, secretaries and administrative assistants have instituted new forms of file review and maintenance activities to fill time made free by word-processing systems.

The Department for Professional Employees of the AFL-CIO has pressed the issue of electronic reviews in campaigns to organize white-collar and professional workers. Such reviews, they claim, lead to the underestimation of personal skills. Citing evidence from West Germany, where more white-collar workers are organized, they argue that managers should develop better protective instincts about the nature of work in the automated workplace.

In the same report, Fritz Hauser, a labor counselor with the West German Embassy in Washington, reported on a study that evaluated the impact of shifts in work: "We also found remarkable changes in qualifications and these were more or less in the erosion of skills and not the improvement of skills."[12] With this kind of evidence, it is not surprising that today's organizing initiatives seem to gather most of their momentum from speculation about the purposes of technology and its effects on the quality of work as perceived by the worker.

Participation: The New Paternalism

Thus far the corporate world has been able to withstand such organizing drives through programs that promise more employee participation and through commitments to retrain workers, especially managers. From the Christmas party to token gifts, such as logo-emblazoned T-shirts and company jewelry, corporate America seeks to instill the idea of family. But participation, to be effective, must link the fate of the employee to that of the corporation. Profit sharing and employee stock ownership are used to foster feelings of participation without involving employees in the most critical aspects of corporate direction— the setting of overall goals and objectives. Financial benefits are much more likely to be shared than power. When the corporation thrives, so do the employees; and when belts have to be tightened, this too must be across the board. The decisions that affect these cycles are, however, still made in the boardroom.

Hewlitt Packard provided an interesting example of the merits of employee participation. With the August 1985 slump in computer sales, Hewlitt Packard was faced with the choice of laying off workers or instituting some cost-reduction strategies. The latter alternative was chosen. A program was set up requiring all employees to take off two days each month without pay. Not only did staff agree to this, but to the surprise of top management, many showed up for work on these unpaid days. Employee loyalty such as this is considered a distinct competitive advantage by Hewlitt Packard.[13] It also reinforces the notion that people work more because they want to and not because they have to.

By tying employee interest to corporate interest, modern corporate culture attempts to create a paternalistic bond. It is a more

refined brand of paternalism than the authoritarianism of a Pullman or a Ford, since it provides psychological and motivational rewards. But this bond still requires a mind set compatible with the corporate environment and an assumption by workers that those directing the corporate enterprise are acting in the best interests of all concerned. Employee perquisites—like health facilities, corporation-sponsored social events, and participative planning—strengthen the relationship. But what is needed to make the bond strongest, a mutual trust contract with mutual guarantees, is not something that today's corporations can provide. The interests of the corporation and of its employees are not the same. As shareholders know, the corporation's goal is survival in the economic system, at whatever price.

The Human Cost

While payroll reduction and downsizing, buyouts, and layoffs may mean salvation to the corporation, they often spell disaster for the worker. Many feel that the loss of self-esteem and the sense of betrayal, the feeling of having been set adrift by the company, are more devastating than personal and family economic adjustments.

There is no reason to believe that employee displacement is always necessary to corporate survival—or even beneficial. Indeed, those who lose their jobs because of merger mania (which sets U.S. corporations apart from those in the rest of the world) may well doubt that their sacrifice left either their former employers or the U.S. economy in better shape.

In spite of the unrealistic employee expectations that many U.S. corporations encourage through programs of integration and participation, many accept little actual responsibility for the short-term or long-term effects of their actions on their workers. Staff realignments, machine-based reviews, changing performance goals, and other stress-provoking actions are making workers increasingly aware of where their interests and those of the corporation diverge. One major area of divergence is the psychological investments workers make in their jobs and their company. It is here that the corporation is most reluctant to accept responsibility. Corporate survival often dictates policies that result in the relocation of jobs to cheaper labor markets.

A recent Department of Labor report on women and office automation, for example, addressed a trend toward offshore clerical work. High-technology transmission systems have made possible the transfer of data entry jobs to third world labor markets, "particularly affecting minorities, older women, and less educated women." [14] In hourly wages, the cost savings to the corporation are impressive. Data entry operators in the United States receive six to twenty dollars per hour; in Barbados, Formosa, or Korea, the average is two to three dollars per hour. It's not surprising that one of the major airlines farms out most of its keypunch operations to Barbados, or that a legal publishing firm transmits material to South Korea, where it is entered into the firm's databank. Data entry is performed there by non-English-speaking operators. Assuming that there is effective quality control, the work gets done. Worker satisfaction, it is safe to assume, is far down in the list of corporate concerns.

Offshore production also offers opportunities to save on health and safety. In their efforts to recruit foreign investment, many third world countries have resisted regulating working conditions. Legal liability is also highly variable, which makes it much easier for operations to be carried out in locations removed from more stringent U.S. standards.

Technological Determinism: The New Corporate Shelter

As we consider change that involves the widespread adoption of high technology, it is seductive to think that the technology itself is causing such changes as unemployment, social disintegration, and class antagonisms. Technological determinism is a very convenient corporate shelter. It excuses companies from the responsibility for social fallout.

First, this attitude somehow allows us to believe that the technology is capable of creating social patterns. According to this argument, numerical control machines determine the division of labor and the methods of work on the shop floor. It is a difficult argument to counter, because some of it is true: the technology of fixed equipment does to some extent set the number and size of shifts, production quotas, and so on. But technology, at least at present, does not make decisions. Management makes the decisions and, we as-

sume, exercises control over worker displacement in terms of both who goes and when.

Second, the ways in which technology is used are also a product of social decision making. Although we often perceive technology as relentlessly efficient, it is human know-how that makes it so. Efficiency experts, programmers, or the machinists they are designed to replace decide the most productive ways to use automated systems. When designing an artificial intelligence program, a model professional worker is carefully studied and tracked so that his or her skills can be captured, codified, and reproduced. Management may gloss over the human side of such planning to give the impression that the technology itself possesses a competence that is, in fact, a systematically programmed substitute for human labor.

Third, when operating from a perspective of technological determinism, there is a tendency to adopt a fatalistic world view. Thus, it becomes possible to believe that technology organizes society so totally that any intervention in the name of morality or ethics is incomprehensible and futile. The system is in place and reproduces itself. The similarity to totalitarianism is obvious, insofar as both remove the system from any context of accountability.

Technological determinism excuses inaction by reinforcing passivity and denying a sense of responsibility. The seemingly mindless workers of Fritz Lang's melodramatic film *Metropolis* are a classic example. They accept their lot, walk with heads bowed, and willingly sacrifice themselves to mammon—at least in the first half of the film.

But human history to date does not seem to favor societies organized according to this principle. The direction of technological development and institutional response is at the heart of any discourse on the future. Commission reports on the year 2000—either in the first wave of the 1960s or the current one—are focused on the steering mechanisms of society: the political, economic, and cultural institutions in which strategies for directing the new technology must be developed.

Although there are not many useful models for these institutions to consider, there are some. IBM, for instance, has buttressed job security with elaborate training and retraining programs. While these might not be generalizable, they are worth evaluating in terms of how they were conceived and developed. Rather than specific so-

lutions, what is most necessary now is an unambiguous acknowledgment of the need for anticipatory planning capable of formulating responses to technological development. This planning must be systematic, controlled, and directed toward reducing the human and social costs.

5

Labor Unions:
Dinosaurs of the 1980s

Organized Labor Lost

Organized labor's responses to the introduction of technology to date have been mainly defensive: agreements ensuring current workforce positions and guaranteed incomes. Differences in strategy depend on the type of industry and the union's negotiating strength.

The "aristocracy" of labor, the AFL-CIO unions, represent workers in the industries—automotive, steel, textiles—where the effects of automation are already being felt in displacement, dislocation, and downsizing. Properly implemented in such industries, high technology not only performs better under hazardous conditions, but makes the human worker more and more redundant. So it is not surprising that these are the industries in which the most antagonistic labor-management relationships have developed.

From the standpoint of management, technological upgrading represents a way to recapture dwindling profits and to keep them by replacing high-wage, high-benefits workers with low-demand, easily supervised robots. To date, labor has had little to no voice in how the new technology will be implemented.

Faced with corporate plans for automation that cite unambiguous productivity statistics, organized labor, in its current weakened state, has been hard pressed to develop solutions that go beyond job protection. Primarily concerned for the current workforce, unions such as the typographers and longshoremen have negotiated protections

for their members. In exchange, they have ceded jobs and protection for the workers of the future. The long-term effects are obvious.

In the less-endangered service and information industries, unions have to date had only marginal success in organizing clerical and other white-collar workers. White-collar workers have historically been antipathetic to unionization, in part because of old blue-collar–white-collar status differences. In these industries, workers also fear loss of bargaining power and workplace autonomy. Unions maintain they can negotiate for improvements in both areas, but the current workforce isn't buying.

Richard Bellouz, a labor market analyst for the Conference Board, assesses it this way: "There is a lot of frustration out there among middle-level workers. . . . They are rapidly forming associations to utilize tactics such as lawsuits and lobbying. A lot of them probably will not pay dues to the AFL-CIO, but that has more to do with snobbishness in not wanting to see themselves mixed up with blue-collar workers."[1]

Unions face other obstacles when they target monoliths such as IBM. These companies have built-in measures to ensure job satisfaction and have corporate strategies to ensure job security. Whether this approach is cooptive, paternalistic, benevolent Japanese management or American corporate resilience, it works. Organizationally it represents a strategy for stability in a complex technological and social environment that is currently being altered by automation.

Organized labor seems powerless to affect loyalties to companies like IBM. For unions to organize in such an environment, they have to be able to make a compelling case. The next five years will be a test for unions as they develop organizing strategies that take the human-relations-oriented corporation into account. Unions must respond to the anticipatory strategies of sophisticated corporations as well as to a residual suspicion that unions are bargaining agents for unproductive, extraneous laborers. It is not enough for union organizers to truthfully remind these workers that most progressive company policies were adopted in nonunion shops only after they had been won elsewhere by organized labor.

In 1977, roughly one private sector wage or salaried worker in five was a union member. Department of Labor statistics on total union membership have not been collected since 1980, but data on employed wage and salaried workers who are union members show

that between 1980 and 1984, union membership declined by 2.7 million to a low of 17.4 million. During the same period, the total number of wage and salaried workers grew by 3.7 million. According to the AFL-CIO Committee on the Evolution of Work, "The proportion of workers who are eligible to join a union and who in fact belong to a union has fallen from close to 45 per cent to under 28 percent since 1954; using the measure of percentage of the entire workplace, the decline has been from 35 per cent to 19 per cent."[2] This decline reflects deindustrialization, relocation of manufacturing, and the rise of service and information industries.

But these trends alone cannot account for so rapid a change. According to Department of Labor data, the number of jobs held by union members in the construction, mining, and manufacturing industries fell by 1.9 million. Concurrently, the number of jobs held by nonunion workers increased by 1.1 million. In the service sector, approximately 5 million jobs were added between 1980 and 1984, as union membership fell by 700,000. Losses were high in the transportation industry, where union activists had anticipated that deregulation would intensify antiunion pressures. Although the rate of unionization in the public sector remained near 36 percent, the loss of 300,000 jobs there resulted in a net reduction in union size of 100,000 members.[3]

What Is Happening? The Challenges

The profile of the typical union member, who is a middle-aged white male in some type of heavy industry, and the changing composition of the workforce, which has more women and minority group members, raises the question of whether unions will be able to appeal to these groups. Senator Paul Simon of Illinois sees this as one of the currently significant nontechnological challenges to unionization. In a reply to a question posed in late 1985 regarding technological unemployment, he said:

> Issues like wage discrimination, continuing education, flexible work schedules, health care costs and child care will become greater concerns. Can unions, which are now 86 percent white and two-thirds men, adjust to these new constituencies with their new concerns? These are the people working in the occupations that are most likely to benefit from unionization.[4]

Even on the crucial technology-related issue of job security, organized labor has an uneven record. Industrial unions have traditionally negotiated to save the jobs of workers with the most seniority. For steelworkers, longshoremen, coal miners, and others in declining industries, negotiations have focused on doing the best for member workers. The United Steelworkers of America, for example, has had to contend with major contractions in the workforce. After seeing the number of steel industry jobs drop from 453,000 in 1979 to 200,000 in 1985, with a forecast of a further 30,000 loss by 1990, the union has called for sacrifices on the part of management and the banks as well as labor.

Industry analysts predicted that in the 1986 contract negotiations, the steelworkers would make some concessions in exchange for job security provisions, stock options, and some degree of participation in management. According to Charles Bradford, a Merrill Lynch steel industry analyst, "I doubt we'll see more wage reductions, but we might see elimination of some work rules, which could result in a 20 percent savings on employment costs at some plants."[5] However, what the country saw was a strike against U.S. Steel—now known as USX—which was the first in twenty-seven years and involved picketing in nine states. Twenty-two thousand union members were affected by the action, which the company called a "strike" and the union a "lockout."[6]

The six-month work stoppage was settled in a contract that demanded concessions on both sides. USX agreed to limit its "contracting out," thereby insuring jobs for union workers. Some estimates are that at some plants almost a third of jobs are subcontracted to nonunion members. In return, the union agreed to wage reductions, cuts in vacation and holiday pay, and changes in health insurance coverage. Some jobs would be dropped through an early retirement plan, whereas others would be saved by agreements to keep mills open. Optimistically, union officials hope that some of the wage reductions will be offset by employee profit sharing. Still, at the end of the strike both sides faced the depressing situation of the steel industry, including predicted price wars.[7]

Steel is a worst-case industry. It has witnessed bankruptcy reorganizations at McLouth and Wheeling-Pittsburgh. An Employee Stock Ownership Plan is being tried at Weirton, and Bethlehem Steel has amended its pension plan for white-collar workers to save

$30 million a year. Industry policy has shifted toward international production and cost savings. USX, for example, will import steel from South Korea for its California finishing plant. Furthermore, the reconfiguration of plants as nonunion shops within the United States has put organized labor on the defensive.

The automobile industry offers another scenario. The United Automobile Workers of America has been confronted with industry-wide reorganizations based on the adoption of high technology and has joined with industry leaders to help preserve jobs.

Union leaders believe it is in the UAW's long-term interests to recognize the inevitability of automation, as it has in the landmark Saturn contract with GM and the UAW-Ford Employee Development and Training Program.[8] Throughout the 1980s, the UAW has negotiated with the industry to allow some room for workers to help develop the technology and preserve some possibility of intervening in its use.

Such negotiations have been marked by the trading of jobs, paid holidays, and benefits for training programs and some degree of participation in management. The union is gambling on the protection of a smaller number of jobs in exchange for production stream-lining and improved labor-management relations. At Ford, where by 1985 the hourly workforce had shrunk, the white-collar staff had been cut by thirty percent, and eight plants were closed, new policies allow workers to stop the production line when faulty parts are being produced. As a result, fewer parts are scrapped—but obviously not fewer workers!

Under the UAW Employee Involvement Program, Ford workers can make suggestions about design and production. Peter Pestillo, Ford's Vice President for Employee Relations, summarized the impact of the program: "These are not dramatic things. . . . But the dozens of employee suggestions we've used has saved us between $300,000 and $700,000 each and given us a better car besides. It adds up."[9]

Social Safety Nets and the New Realism

The UAW realizes that the large numbers of lost jobs are not going to reappear. Realistically, it has put the problem of displacement in

the larger economic context. In doing so, the union calls for a national labor policy that protects the rights of workers in shifts to newer industries. The 27th UAW Constitutional Convention, held in May 1983, adopted the following resolution regarding a more comprehensive national approach to displacement:

> The nation needs a policy to minimize undesirable economic dislocation. With the pace of economic change already rapid, significant economic changes that do not serve a wider public purpose should be restricted. Comprehensive plant closing legislation will be necessary . . . there should be changes in the tax code to eliminate provisions which encourage plant runaways and relocations. A plant closing law must also include a code of corporate conduct in plant shutdown and relocation situations. . . . We also need a bill of rights for displaced workers and dislocated workers. Victims of economic change need a truly adequate "social safety net" and various forms of assistance: job search, relocation, and retraining.[10]

Such a social safety net is necessary, the union maintains, because "training is no solution when lack of jobs is the actual problem."[11]

Accepting the realities of displacement and unemployment poses a problem for unions in all industries that are being radically restructured. Convention resolutions have sounded the call for a new realism on the part of organized labor. This realism accepts the declines in manufacturing and construction industries as inevitable, the product of structural changes in the economy. Union leaders also acknowledge the realities of the shifts in jobs to geographical areas where levels of organization are low. Part-time work, the proliferation of job titles that blur the distinctions between management and labor, and the reduced number of opportunities for workers in areas where jobs are scarce are all realities with which they must contend.

In 1985, the AFL-CIO Committee on the Evolution of Work produced a report entitled "The Changing Situation of Workers and Their Unions," which acknowledges the hard fact that "no serious observer denies that unions have played and continue to play a civilizing, humanizing, and democratizing role in American life." Yet, "despite their accomplishments, unions find themselves behind the pace of change."[12] The report stirred considerable debate within the union community. It took a hard look at the factors noted above and the pressure exerted by the movement of jobs to cheaper international

labor markets. On the issue of technology, it also accedes to the realities of displacement: "Technological advances have eliminated scores of jobs, altered the requirements of an equal number, and created entirely new jobs." In response, the labor movement has had success in "improving wages and working conditions." This success, it is argued, "has had its effect on what workers see as their right, on what workers seek in further improvements, and on what employers recognize as the minimum conditions they must offer."[13]

The extent to which workers identify these gains as the product of union representation is unclear. The Committee on the Evolution of Work acknowledged that American workers are "ambivalent in their attitudes toward unions." The committee reviewed surveys on public attitudes toward unions over the last twenty-five years and found that:

> Over 75 percent of all workers—and over 75 percent of non-union workers—state that they agree that unions in general improve the wages and working conditions of workers. Over 80 percent of all workers agree that unions are needed so that the legitimate complaints of workers can be heard. Yet when asked to assess the effects of organization on their present employer, 53 percent of non-union workers state that wages and fringe benefits would not improve and 74 percent state that job security would not improve.[14]

Confronted with American workers' needs for independence in the workplace, the resilience of the work ethic, and the results of survey data which indicate that nonunion workers identify unions as monopolistic and nondemocratic, the AFL-CIO committee concluded that American unions need to improve their image. To have an impact in employment areas that are traditionally not organized, unions will have to demonstrate that they are neither inflexible nor oblivious to the economic constraints industry faces. The information and service industries will be a testing ground.[15]

One Challenge: Organizing the Information and High-Tech Sectors

Information and service industries pose different challenges for organized labor than do manufacturing industries. Here, the question

is whether organized labor can respond to the needs of a workforce with a great many part-time and upwardly mobile workers who aspire to white-collar and professional status and who have little if any experience with organized labor.

According to the AFL-CIO Committee on the Evolution of Work, ninety-five percent of all private sector employers actively resist unionization, and seventy-five percent hire "labor management" firms to avoid unionization.[16] In a 1985 survey of young workers across the country, *Fortune* writer Michael Brody found that "They are not hostile to unions in principle but most seem indifferent to them."[17] Resistance on the part of corporations, combined with employee reluctance to affiliate with unions, presents a strong challenge. Yet unions representing workers in information and service industries have amassed considerable evidence that collective bargaining would insure better working conditions and job security.

These considerations are particularly germane for lower-echelon clerical, technical, and service workers, who are more likely to be unprotected by employee contracts, and for part-time workers with few or no benefits. Since a significant part of this population is female, pay equity also becomes a possible bargaining issue.

Unless organized labor makes strong inroads into the information and service industries, it will continue to suffer shrinking membership, with correspondingly decreasing power and ability to influence public policy. While rising membership in one union in one industry does not directly strengthen another union in another industry, the more workers who are organized, the unions argue, the more powerful will be the voice of labor. As the erosion of jobs in heavy industry accelerates (the parallel with declines in the agricultural workforce in the early twentieth century comes to mind here), the need to organize in the fastest-growing sector of the economy becomes even more critical. So when the Communication Workers of America announced in 1985 that IBM was its next organizing target, the union was taking aim at the employment opportunities of the future, and preparing to tackle the recent blurring of distinctions between white- and blue-collar work.[18]

The London-based International Labor Organization has reported that only about 10,000 of IBM's worldwide workforce of 395,000 belongs to any type of labor organization and that none of IBM's U.S. employees belong to a trade union.[19] In its campaign to

organize IBM, CWA faces an impressive corporate image. IBM has a reputation of being a very good employer. Its employee training programs are cited as models for other companies, and historically there has been a commitment to workforce stability. (IBM's annual staff turnover rate is less than four percent.) Fringe benefits and corporate amenities are considered generous, and employee interests have historically been protected by management. Yet when it comes to worker participation in job definition, Morton Bahr of CWA argues that "workers have no say about their jobs."[20]

Organizing blue-collar workers involved clear-cut issues of working conditions, job security, and workplace safety, but at IBM blue-collar–white-collar distinctions are difficult to make.

Loss of the distinctions between types of work is occurring everywhere. A *Fortune* survey of young workers aged eighteen to thirty-five, which included workers in factories, offices, bars, and unemployment lines, found that: "Many seem not even to recognize the hallowed term blue collar, so much has the workplace changed in the 1970's and 1980's. Indeed, many have made their way up and out of low-skilled, low-paying white collar jobs at the checkout counter or filing cabinet to more highly skilled, higher-paying blue collar work."[21] Whether or not the *Fortune* survey can be generalized, it does help us understand some of the attitudinal obstacles that unions encounter in high-tech workplaces.

Unions have traditionally maintained a commitment to solidarity and collective bargaining, while achievement in the white-collar and professional world is perceived as highly individualistic. The division between blue- and white-collar work has a psychological component in the traditional association of unions in the United States with the working class and of white-collar work as the escape from that class. As Richard Sennett and Jonathan Cobb pointed out in *The Hidden Injuries of Class*, it was not always easy to leave behind class values as a worker made the shift from manual-production work to knowledge-information work. At the same time, an upwardly mobile person might have a hard time with membership in unions since it blurs the progress that the shift in collars represents.[22]

The adversarial roles that unions and management have chosen is another legacy that handicaps union initiatives in high-tech industries. Most corporations have adopted some type of human relations strategy that mitigates conflict. Quality work circles and

employee stock ownership options are examples. In response to the loss of jobs during the recessions of the 1970s and early 1980s, the industrial unions have assumed a somewhat more cooperative role, working with management to develop training programs and open avenues of employee participation. To some degree, the roles that unions will have to play if they are to become a bargaining force in the information industries are now being developed in heavy industry, where both companies and unions recognized the need to adapt or die.

A Second Challenge: Reorganizing Manufacturing

In addition to organizing the high-technology information industries, unions must also find ways to attract the large numbers of unorganized workers in manufacturing industries. The 1985 AFL-CIO report on the evolution of work argued that, since goods-producing industries still employ 26 percent of the working population, organizing efforts there must be maintained. The report specifically targets substantial numbers of unorganized workers in major industries:

more than 50 percent of the metal, machine, and electrical workers;

69 percent of chemical, oil, rubber, plastics, and glass workers;

64 percent of wood, paper, and furniture workers;

67 percent of food processing workers.[23]

Some of the reasons for these large numbers are: regional shifts that restructured industries on a nonunion basis; the rise of new unorganized industries; and an increase in nonunion companies. According to the AFL-CIO, at least 27 million, or 26 percent, of nonunionized workers are former union members who changed jobs and now work in nonunion shops. These workers are prime targets, since they are familiar with the benefits of unionization and, presumably, comfortable in the ranks.

Since the late 1970s, unions have seen the steady erosion of benefits gained in earier contracts. The 1979 United Auto Workers–Chrysler agreement is an example, because it included both pay cuts and the loss of paid holidays. Wage reductions, freezes, delayed wage

increases, and lower wages for new hires followed in industry after industry. A Federal Reserve Board study has shown that between 1981 and the fourth quarter of 1984, roughly 4 million workers accepted contracts that included wage freezes or cuts. [24] Such patterns of concessions can be found in the auto, steel, trucking, airline, aerospace, meatpacking, and food industries. The inclusion of high-tech and service industries to this list makes it clear that not only rust belt workers face difficult times when they renegotiate wage and benefits packages.

In many industries, and even in such high-tech companies as AT&T, a double standard for wages (known as the "two-tier wage agreement") has become increasingly common as a means of securing job protection for unionized workers. One labor historian has said: "The two tier wage system has proven to be one of the most corrosive influences on unions' ability to maintain levels of organization." If we consider the Packard Electric "job security pact" between the International Brotherhood of Electricians and GM, we can see why. It included, in exchange for "lifetime" job security guarantees, a two-tier wage provision that brings new hires in at about forty percent of the compensation rate for current job-holders.

A report published by the Conference Board in March 1985 gives an indication of what unions will face in negotiations on cost cutting. Citing results of a follow-up survey of board members covering the period 1978–1983, Audrey Freedman reported: "There has been an extensive shift in the five-year period toward management demand for 'give-backs' in the non-wage and fringe area." According to 369 companies who responded to this part of the survey, the following cost-cutting objectives had highest priority:

"flexibility in assignment of employees" (i.e., changes in work rules);

reduction of health care and insurance provisions;

delaying, reducing, or eliminating cost-of-living adjustments;

layoff and recall procedures, with a weakening of seniority provisions;

income security (severance pay, supplementary employment benefits);

time off with pay (vacation and holiday time, sick leave, wash-up, rest and lunch periods);

reduction of contract periods to one year.[25]

To date, health benefits have been hardest hit, but pension provisions are also being amended. According to *Economic Notes*, published by the Labor Research Association, 63 percent of all contracts negotiated during the first six months of 1985 contained revised pension plans.[26]

Fringe benefits are a controversial bargaining area. American health care costs are high, and without these benefits most wage-earners would be unable to afford proper care. By contrast, critics argue that fringe benefit packages are often padded. Yet a 1984 study by the Congressional Research Service looked at average benefits throughout the world in 1981, with these results: in terms of comparative value (value of benefits calculated as a percentage of compensation for actual time worked), Italy did best by its workers, with a value of 83 percent. Japan and West Germany were next, with 75 percent and 71 percent, respectively. The United Kingdom followed at 38.3 percent, and then came the United States at 37 percent. U.S. unions, despite this comparative disadvantage, are increasingly put in the position of having to choose between fringe benefits and job security for workers, and they have opted for the latter.[27]

More Challenges: The Antiunion Attitude

Unions face another challenge when they go to the bargaining table. This involves a sharply defined attitude on the part of the administration in Washington that unions have gone too far. Morton Bahr of CWA sees President Reagan's response to the PATCO (air controllers) strike of 1981 as a clear expression of this. It "sent a signal to the Board Room—'Get tough with unions, I showed you how.' "[28]

AFL-CIO unions report facing employers who are highly resistant to unionization. They argue that the Wagner Act provisions are becoming impotent and that the Labor Board is "inert." Unions question how a National Labor Relations Board chairman who makes statements like "collective bargaining frequently means . . . the destruction of individual freedom and the destruction of the market-

place," and "the price we have paid is the loss of entire industries and the crippling of others," can administer labor law fairly. Combined with the pressure of a structural shortage of at least 4 million jobs through the 1980s, antiunion policies on the part of the administration make organizing drives difficult.[29]

Workers' perceptions of unions make organization no easier. In repeated surveys, a substantial number of both unionized and non-unionized workers have expressed the view that unions are undemocratic. The AFL-CIO Committee on the Evolution of Work cites studies which indicate that "among the population as a whole, 50 percent state that they believe that most union leaders no longer represent the workers in their unions." Nonunion workers indicated that unions "force members to go along with decisions they don't like," that "union leaders, not membership, decide who will go on strike," that "unions stifle individual initiative," and that unions "fight change."[30] While union members express far more positive views, these attitudes present a formidable challenge to organized labor.

Part of the new realism of unions involves confronting the stereotype of themselves as monopolistic. In their defense, they present the benefits that organization has brought to labor. Job security and more say in work are critical evidence. So is the social safety net unions have created through fringe benefit packages—medical insurance, pensions, child care, and the like.

According to Freeman and Medoff, authors of *What Do Unions Do?*, unions also voice social needs and concerns, and can (when the system functions well) contribute to greater participation in the democratic process. Collective representation, they maintain, provides the safety of numbers for those who speak out. It also permits bargaining on such issues as safety conditions or benefits that insure the rights of the majority.[31]

Currently, we accept as normal a seven percent level of unemployment and up to 8 million permanently displaced workers, in spite of what it costs the unemployed in self-esteem and material disadvantage. Unions in their advocacy role argue that provision must be made to insure that work is available for all, even if some collective means of underwriting it is required.

In their role as organized labor's voice, unions still need to convince most workers that they are capable of representing their best

interests. This is a tough challenge. But unions have the loyalty of their membership and an impressive if uneven historical record. They also have the advantages of organizations in place and of established relationships with policymakers.

As this chapter goes to press, some new patterns are beginning to emerge which indicate that organized labor's drive into the high-tech workplace is beginning to show results. In 1986, union membership fell by 21,000 to 17.5 percent of the workforce. When compared with membership declines of 344,000 in 1985 and 377,000 in 1984, this relatively small loss is interpreted as evidence of organizational gains among white-collar workers and women.[32] While these figures represent only a one-year change, they do indicate resilience in the union movement. When combined with the parallel increase in the number of employee associations, they reflect growing recognition of a need for some mechanism of collective bargaining. Although workers clearly discern this need, the questions of who will represent them and to what end remain open.

6

Unemployment, the Underclass, and the Ideology of Retraining

W HATEVER responsibility society has for the structurally unemployed, the issue of job opportunities for the underclass
is always raised when training and retraining are debated. In part,
programs like WIN (Work Incentive Program, 1967) and CETA
(Comprehensive Employment Training Act) were created in response to the employment needs of those defined as the permanent
poor. Providing employment skills, including the self-discipline required to hold a steady job, has long been considered one means of
reducing the threat that a permanently unemployed adult population
poses to the social order. The need for such programs—if not for
actual jobs—has been demonstrated by urban riots, inner-city crime,
and family violence. Whether structurally unemployed men and
women *can* be trained, and if so who will employ them, are
perennial issues. Shrinking federal aid and local government
payrolls along with the suburbanization of lower-level clerical jobs
are further reducing the limited employment opportunities available
to the underclass.

Public and private, training and retraining programs are designed
to balance capabilities, skills, and available jobs. The more successful
programs operate on the assumption that workers can be equipped
to meet the demands of a changing labor market and that secure jobs
await them upon completion. To succeed, such programs must contend with major challenges: poor basic skills, technical obsolescence,

and overall economic shifts, such as those mentioned previously, as well as a decline in the number of manufacturing jobs. Inadequate school systems, community attitudes toward employment, and inadequate or inaccurate labor market information also raise the odds against success.

Scientific, professional, and technological workers, regardless of status, are in a strong position to argue for up-to-the-minute training. It is the key to their performance and productivity. The underclass, however, are less aware of the gaps in their skills and have less voice in redressing them.

Work programs that include some type of on-the-job training have always been controversial. The New Deal programs, such as the Works Progress Administration (WPA) and the Civilian Conservation Corps (CCC), raised serious debate as to whether government should intervene in this manner, even when unemployment reached unprecedented levels. Widespread criticism of the 1973 Comprehensive Employment Training Act (CETA) program for its failure to teach workers useful skills raised more serious questions about federal support. With the passage of the Job Training Partnership Act (JTPA) in 1982, the Reagan administration committed itself to training and retraining with greater involvement by local private industry councils (PICs) and local elected officials. Advocates of the program see JTPA as immune to the kind of criticism programs like CETA received, since it requires private sector participation in both the planning and the administration of the programs. By involving state and local government in setting policy, the architects of JTPA hope to respond to regional and community needs in a manner similar to block grant programs. JTPA establishes disadvantaged and displaced workers as target populations and makes programs to furnish these workers with marketable skills a community responsibility.

Whether or not JTPA indicates the federal approach to training and retraining in the 1980s, both its critics and supporters agree that it alone is not a sufficient response to technological unemployment. No one is more aware of this than workers now threatened by displacement. Their fears have prompted labor negotiators to press for benefit packages that include mandatory contributions for training/ retraining programs and job guarantees. The UAW Saturn contract and the recent Chrysler contract both include such provisions. Em-

ployers and employees share the cost of these training programs, with employee contributions collected through payroll deductions.

Professionals, white-collar workers, and others not covered by collective bargaining contracts, however, are most often dependent on corporate largess for training. Model programs have been developed by GE, IBM, and Xerox to upgrade and enhance employee skills. But such programs are corporation-specific and seldom include job guarantees. In Europe, the training needs of these groups are being addressed at the national level. Sweden, France, and West Germany support training and retraining, as well as programs to redistribute work, that include white-collar workers and professionals. [1]

Both the retraining needs and employment potential of professionals and white-collar workers are substantially different from the needs and potential of the structurally unemployed. Most of these jobless Americans belong to that portion of the population described by Ken Auletta in his book, *The Underclass.* [2] Looking at the underclass in 1982, when it numbered approximately 9 million people, Auletta divided them into four major groupings:

1. "The passive poor," long-term welfare recipients who are often part of a lineage of intergenerational poverty.
2. "The hostile street criminals," responsible for much of the fear of crime and violence within inner cities, often drug addicts and high school dropouts. Their criminal activities mean that, while they are not necessarily poor, their income is produced by illegal means.
3. "The hustlers," who, unlike street criminals, do not usually commit violent crimes, but earn an income through the underground economy.
4. "The traumatized," the homeless, drunks, and drifters—many of whom have been discharged from mental institutions. [3]

Many critics of welfare argue against providing programs for these groups, since the probability of success is far below that of dislocated workers with work histories. This rationale, however, does not allow us to ignore the young people Gordon Berlin of the

Ford Foundation identifies as most severely at risk in the labor market: urban minority youth, rural youth, and potential high school dropouts.[4] They are not yet part of the underclass. But without intervention, that is what they are destined to become.

With literacy, life skills training, and a basic sense of the workplace, alternatives for these groups are possible. A successful youth program in Chicago demonstrated during 1981 and 1982 that the right kind of job training can work. A local Private Industry Council, recognizing a shortage of word-processing operators, hired a consulting firm to train 100 disadvantaged CETA (later JTPA) applicants in basic skills and word processing. In the course of the program, seventy percent of the participants found employment as word processors.[5].

There is good reason to question the usefulness of the notion of an underclass—and Auletta's formulation of its membership—to the question of job training. Sar A. Levitan and Clifford M. Johnson, citing surveys conducted at the University of Michigan, argue that households move in and out of poverty due to changes that include economic conditions and demographics. As a result, most of these families experience a mixture of work and welfare in the course of their lives.[6]

Whether programs are designed for displaced workers, welfare mothers, or underclass youth, the most important factor is availability of jobs. Accurate assessment of local labor market conditions is essential to program design. So is a realistic consideration of special needs.

In general, basic skills, including literacy, writing, and mathematics, are essential components of training and retraining programs. A 1982 survey by the Center for Public Resources of 184 businesses and 123 school systems nationwide validates these criteria. Sixty-five percent of the companies that responded reported that basic skill deficiencies limited the job advancement opportunities for high school graduates. Thirty percent said that secretaries have difficulty reading at the level required by their jobs, and fifty percent reported managers and supervisors unable to write error-free paragraphs. Fifty percent also reported that bookkeepers and other skilled employees were unable to use decimals and fractions.[7]

Just as disturbing as these survey results are some early 1980s statistics: 13 percent of white, 43 percent of black, and 56 percent

of Hispanic seventeen-year-olds are functionally illiterate.[8] Between 40 and 50 percent of students in urban areas have serious reading problems. And this is the generation that is coming into the labor market of the 1980s and 1990s.

These problems also appear among displaced workers. Freda Rutherford of the successful Downriver Community Conference in Lansing, Michigan, reports on how displaced workers are grouped according to the types of services they need to conduct an effective job search. The four groups range from the job-ready to those requiring a skill retraining program. For the job-ready group, all that is needed is available employment. The second group requires help in such areas as job search techniques, résumé writing, and interviewing skills. The third group—which is larger than had been expected—requires remediation in reading, writing, and mathematics *before* they are considered ready for private or public retraining programs. Members of the fourth group are assessed as ready for retraining.[9]

Basic skills programs with well-defined performance measures are therefore a key component of training. The Center for Public Resources survey, participants at the Northeast-Midwest Institute hearings, and administrators interviewed at the New York City PIC all agree. However, some combination of training, retraining, and educational reform to assure necessary job skills is not enough. Administrators and evaluators of manpower programs repeatedly assert that job availability is the pivotal issue, and it is seldom systemically addressed.

The Evolution of Government Responsibility

In the history of American labor, the Great Depression marks the point at which responsibility for unemployment became a policy issue. Earlier, sick or laidoff workers, if their resources were exhausted, relied on their families or were forced to turn to local charities. Around the turn of the century, national movements emerged in Europe pressing for some type of public welfare system. Belgium adopted the first unemployment insurance plan in 1911.[10] In the United States, things moved more slowly. Writing in *Good Jobs, Bad Jobs, No Jobs*, Eli Ginzburg states that in 1932 the Executive Council of the American Federation of Labor "voted against governmental

assumption of responsibility for unemployment insurance on the grounds that this problem should be solved by management and labor working together without governmental interference."[11] Ginzburg notes that the federal government's response to the widespread unemployment of the 1930s set parameters for national manpower policy that are still intact. These include: "income transfer programs, macroeconomic policies aimed at maintaining high levels of employment, and developing special training and employment opportunities for the disadvantaged."[12]

The New Deal work programs laid the foundation for such intervention, and included the WPA (Works Progress Administration), the CCC (Civilian Conservation Corps), and the FSA (Farm Security Administration). Together, they provided immediate relief for large numbers of the unemployed through projects ranging from Post Office murals to the Tennessee Valley Authority—and the effects were both immediate and effective. WPA at one time employed as many as 3 million people.

The New Deal left behind it the landmark social security legislation and the precedent of federal intervention in the labor market through direct sponsorship of work programs, including skills training. During and after World War II, manpower policy was shaped by a wartime economy in which everyone worked. The needs of the military for a skilled workforce and the GI Education Bill helped to improve the overall skill of the male workforce. The shift toward technology and science that was the key to the development of computers drew momentum from the Cold War and the competitiveness sparked by Sputnik. With the introduction of progressively more sophisticated automation technologies in the 1950s, concern about the need for some federal response to potential wide-ranging unemployment reemerged.

The 1960s brought an expansion of the federal government's role in addressing employment issues. The combination of a national social conscience regarding poverty and a sympathetic alliance of Congress and the Kennedy and Johnson administrations moved the government further into the role of a skills training provider. The 1961 Area Redevelopment Act (ARA) established skills training programs as one means of addressing a depressed local economy. It was enacted on the premise that pools of skilled labor would increase the

likelihood of an industry locating in an area of high unemployment if training was available and carried out at federal expense.

Retraining, Workforce, and Other Band-Aid Approaches

In 1962, the first legislation was passed to address the displacement of workers due to automation. The Manpower Development Training Act (MDTA) authorized funds for retraining skilled workers displaced by technological upgrading of plants. MDTA legislation provided for occupational training in a classroom setting and involved testing, counseling, selection, job development, referral, and job placement. An on-the-job training (OJT) component involved instruction and supervised work at the job site and allowed for classroom training at either a vocational education institution or the employer's location. National unions, multiplant companies, trade associations, and public agencies were all eligible contractors.

But the MDTA programs were never fully tested. Rates of automation proved to have been overestimated, and skilled workers were able to make the transition to different industries without the need for the extensive retraining that had been anticipated.[13]

The real problem, as usual, centered on the unskilled. MDTA specified that a proportion of the target population should include "disadvantaged" unemployed workers. The Economic Opportunity Act of 1964 firmed up the federal commitment to challenging the negative effects of living at or below the poverty level. Jobs Corps and the Comprehensive Employment Program had as special targets disadvantaged urban youth.

Concurrent with the development of jobs programs was an expansion of federal and state welfare programs. By the late 1960s, expanding welfare rolls and rising fears that dependence on income transfers bred a "culture of poverty" prompted the creation of programs to provide employment for welfare recipients. By the mid-to-late 1970s, many states, including New York and California, were experimenting with "workfare."

Workfare, although it varies by state, basically requires recipients of public assistance to work for their grants. An equation is established between the amount an eligible household receives and an

hourly value for a targeted job. While this sets up an exchange—work for aid—it also creates some confusion about appropriate employment for welfare populations. Work for local government does not equal work in private industry, since the skills acquired in government work may not be transferable.

Critics of workfare also point out that some clients in these programs can become more dependent on welfare. They wrongly assume that their grant check is a paycheck and that they have a job with the agency to which they have been assigned. Rather than concentrating on a job search for a position in the private sector, they begin to settle in and form workplace attachments.[14] In addition, workfare raises concern among municipal unions, since hard-pressed local governments often find ways to substitute grant-paid workfare recipients for full-time union workers.

The 1982 Job Training Partnership Act targeted welfare recipients, unemployed youth, and displaced workers. Title III, the Dislocated Worker Program, was designed to "assist workers who have lost or are at risk of losing their jobs because of plant closings and massive layoffs caused by world competition and technologcal change." Its focus on technological unemployment makes it a direct descendant of Title III of the MDTA OJT program.

According to the legislation, the states have almost complete authority over how the program is targeted, how resources are distributed, and what services will be provided. The benefits of state control range from a greater familiarity with providers of training to the more effective use of discretionary funds. These funds can be awarded, based on performance measures, to encourage competition among different service delivery areas and among program sponsors within these areas. The goal of the legislation is to tie job training programs to local development planning, thus shifting responsibility from the federal to the local level.

An independent assessment of the initial operating period of JTPA from October 1, 1983, to June 30, 1984, conducted by Gary Walker, Hilary Feldstein, and Katherine Solow found that placement rates were higher than expected: 70 percent versus a 58 percent standard for adults, and 64 percent versus a federal standard of 41 percent for youths. However, these results are preliminary and must be tempered by the finding that JTPA's targeting requirements for those "most in need" were in general not satisfied. This was especially

true for youths, dropouts, and other high-need individuals. Furthermore, the study pointed out the need for more effective cooperation between state and local governments.[15]

In contrast to the experience of JTPA was the Emergency Jobs Appropriations Act of 1983. The act, passed two months after the national unemployment rate peaked at 11.4 percent, established a $9 billion program that a General Accounting Office report in January 1987 judged to be a failure. In raw terms, 35,000 jobs were created at a cost of $88,571 each.

The reasons for the failure offer some sobering insights into the complexity of creating a response to unemployment. According to the GAO: "Most funds made available by the act were not spent quickly and relatively few jobs were provided when they were most needed by the economy." Furthermore, little evidence was found that hiring the long-term unemployed, the target group for the act, was emphasized. The GAO report argued: "Had the act emphasized programs and activities that could have spent funds quickly, before the economy began to recover, more jobs would have been provided when jobs were most needed following the recession."[16] It should be obvious to planners that a quick-fix program that throws money at a problem will get bad marks, particularly if it is run by government. Thus, the program probably did more harm than good.

Good Programs, Bad Programs

Sensitivity to local conditions has long been a key to successful training programs. Eli Ginzberg, in his *Good Jobs, Bad Jobs, No Jobs* (1979), lists six variables critical to the success of adequately funded manpower programs:

1. The ability and availability of program administrators and project leadership at the community level.
2. The political climate and the degree of support from vested interest groups.
3. The quality of the clientele served.
4. The adequacy of supportive services offered.
5. The economic circumstances in which the program or project operates.

6. The timing and preparation for the mounting of efforts.[17]

Because manufacturing jobs are fast disappearing in New York, the Private Industry Council managers in New York City have designed a package that involves counseling, educational skills, and skills retraining. It is intended to provide a prototype for delivering services to workers in phased-out industries. An employment assistance center to be established at a plant in Queens will be considered by the workers as the "nexus of services." The Public Development Corporation and the Financial Services Corporation, as part of New York City's economic development infrastructure, will work with the local development corporation to identify potential jobs. The Board of Education has agreed to establish a literacy center to provide remedial classes. The PIC will contract for retraining services and provide job search and counseling skills, while the relocating company will provide such in-kind assistance as facilities for training, job contacts, and the like. The New York City PIC has focused on industry-specific skills, including security operations, clerical skills for the insurance industry, copier repair skills, and other service skills.

The people who will participate in this project will not yet have exhausted their unemployment benefits or other financial resources, which makes the project a departure from the norm. Although national evaluations of displaced worker training projects have found that these programs can usually be most effective about six months after layoff, the New York PIC has decided to try to intervene earlier. The six-month hiatus is believed to increase program effectiveness, because workers by that time are beginning to realize that unemployment benefits will expire and the hoped-for call back is not going to come. If the New York project proves successful, it may lend credence to the notion that an anticipatory program, responsive to the needs of the displaced, should be enacted.

Interesting questions are raised in all of the discussions, studies, reports, debates, and seminars on the postindustrial society about the lack of attention to what programs actually teach. Program focus seems to be on what is politically expedient—something that gives the illusion of action. Unfortunately, illusion and the inevitable harsh evaluation go hand-in-hand. There may be no solution to job loss

and job degradation as we move to a high-tech society. A permanent program to train, retrain, and place the workers who are caught in the transition, however, is a measure of a just society.

7

The Labor/Economic Crisis

THE predictions are that job displacement caused by high technology will, in the worst case, produce massive unemployment. In the best case, it will reduce the earning power of the average worker sufficiently to create an expanding underclass composed of the unemployed and the underemployed. We prepared for this kind of havoc in the labor force once before. Fear of automation during the 1950s and 1960s was based on predictions of similar large-scale displacement. Assembly line workers feared losing their jobs to robots, and clerks and typists anticipated replacement by computers. In response to this widespread "automation scare," federal training programs, including those funded by the Manpower Development and Training Act, were launched with the explicit purpose of retraining technologically displaced industrial workers.

Workers in the 1950s and 1960s, however, did *not* suffer significant displacement. With the expansion of information industries, the economic recovery of the 1960s, and defense spending for the Vietnam War, the status of most workers improved. Computers and the accompanying communication technologies evolved more slowly than had been projected. Once chip technology allowed miniaturization, however, costs for high-technology computer equipment began to decline. All this contributed to the overall growth of the economy and to new jobs. Now, with the maturing of computer and communication technologies, deindustrialization, and the eclipse of heavy industry, plus serious international competition, we are finally seeing some of these old predictions realized.

The Shock of Losing

Newspapers sound daily alarms about expanding trade deficits (in 1986 the figure was somewhere in the region of $170 billion) and low productivity. In 1985, the United States became a debtor nation and the debt was accelerated by currency market shifts and a negative trade balance. By 1986, its debt had become the largest in history.[1] According to economist Lester Thurow, U.S. productivity—that is, output per hour of work—has already been surpassed by Germany and France.[2] Human interest stories regularly focus on plant closings, downsizing of middle management, and the shrinking of the blue-collar middle class.

Although unemployment stabilized at seven percent at the end of 1985, this figure does not truly reflect the number of Americans who cannot find jobs. Those who have left the labor market—the discouraged workers—are not counted, nor are the residually unemployed, those who experience the chronic effects of unemployment. These people are not the victims of high technology and automation. Some are not able to work because of physical and/or psychological handicaps, and others have been so much a part of a culture of poverty that they have not learned how to perform within the labor force.

For those who want to work and who have the skills, technological displacement is not likely to mean unemployment so much as lesser employment—a shift to what work is available—generally at lower pay. While the number of very skilled professional and managerial jobs at the high-paying end of the job spectrum will be increasing, so will the number of minimally skilled service positions at the low-paying end. It is the middle that shrinks, and workers displaced from middle-level positions are far more likely to drop down than scramble up.

Even now, the prospects for those who lose their jobs because of automation are grim, at least in the short term. Displaced workers and their families who are forced to lower their standard of living by as much as fifty percent face some hard choices. The impact on the quality of family life is severe, but when the numbers are large enough, this impact is also felt by the economy because these families curtail the consumption that makes the economy work. The effect

on the marketplace was already measurable enough in 1983 to be reported in *Fortune*.[3]

Nevertheless, public attention and political concern have not focused on the downshift of displaced workers as much as on the possibility of widespread unemployment resulting from the inefficiency of the old technology—not from the efficiency of the new. James B. Hunt, governor of North Carolina and chairman of the National Task Force on Education for Economic Growth, 1985, summarizes it this way:

> The possibility that other nations may outstrip us in inventiveness and productivity is suddenly troubling Americans. Communities all over the United States are depressingly familiar now with what the experts call technological, or structural, unemployment; joblessness that occurs because our workers, our factories and our techniques are suddenly obsolete. To many Americans, technological change today seems a dark and threatening force, rather than a bright confirmation of our national genius.[4]

In *The Zero-Sum Solution: Building a World-Class American Economy* (1985), Lester Thurow gives an overview of why the United States now lags behind in productivity:

> In many ways what is needed is the moral equivalent of defeat. No society undertakes the painful efforts necessary for reconstruction unless it understands that it has been defeated—that it cannot continue as it has been going, and must rebuild in new ways or lose its traditional economic position. . . . The real advantage flowing from defeat was a social advantage—a willingness to change old ways for new ways, a willingness to give up old pretensions. Having been defeated, one cannot maintain that the old ways were good ways. The old ways have just failed. They must be replaced.[5]

Defeat is a strong word—perhaps too strong—to use to assess American productivity and its competitiveness in an international marketplace. Yet Thurow is right to try to shock political and economic leaders out of their complacency.

Changes in standards of living are a key indicator of a society's ability to provide for its citizens. Studies of displaced workers, in-

cluding middle managers, show a steady reduction of the middle class. Measuring the impact and extent of this reduction is one of the most complicated side issues of technological unemployment. It is often difficult to tell who is unemployed because of industrial obsolescence, who because of new technology, and who through "reorganization."

One measure of the phenomenon is the recent decline of the percentage of men holding middle-income jobs. From 1976 to 1983, the percentage of jobs that fell into the middle-income range—between 75 and 125 percent of median male earnings—declined from 23.4 to 20.4 percent of the male workforce. From 1977 to 1983, the percentage of middle-income households dropped from 24.0 to 23.2 percent of the population.[6]

Thurow and others warn of the social threat posed by the loss of a heterogeneous middle class composed of both blue- and white-collar workers. The decline of the middle class could result in withdrawal from participation in the democratic process. A firmly held tenet of American political theory is that the middle class provides the cohesion to sustain a democracy. Historically, its members have supported the democratic process by membership in voluntary associations and political parties. Without their investment and participation in the system, the fate of our democratic institutions could be threatened and our political stability put at risk.[7]

Social scientists have long identified the ability to maintain a particular lifestyle as a key source of identifying for Americans. Being unable to compete produces feelings of marginality that can lead either to internalization of the loss of participation and withdrawal from the community or to mobilization and the possibility of organizing politically and economically to make changes.

The Ripple Effect:
Some Hidden Costs

There are some clear underlying assumptions concerning the impact on the democratic process that results from technological unemployment. One is that prolonged unemployment at seven or eight percent has a significant impact on all the major institutions of a society, not just on the economy and the labor market. This assumption has guided most of the social welfare legislation of the

twentieth century. Another assumption is that American society wants to remain committed to equity, the democratic process, and the promise of upward mobility. The last of these, the belief that hard work and effort can earn entry to the middle class, has been critical for the development of American society. Now, it seems, that belief is being shattered by the impact of technological unemployment and underemployment.

Structural unemployment belies the idea that people can work their way up and that hard work is its own reward. By accepting an economic definition of structural unemployment, "the technical name for the loss of jobs when changing technology makes current held skills obsolete," we can begin to see how technology affects unemployment.[8] As technology reshapes the workplace, it influences the types of jobs that are available and the skills that are required. As Leontief has pointed out, the mechanization of agriculture radically reduced employment prospects in farming while enhancing productivity. What we have argued in preceding chapters is that computerized technologies and the management policies that accompany them pose a similar, but graver, challenge. The response to those challenges will not necessarily emerge from the economic institutions that create them.

Look, for example, at the banking industry, which has historically maintained a reputation of taking care of employees. Even in the industry's terms, it was considered paternalistic. The orientation of bank training programs in the last ten years, however, marks a real shift in attitude. The employer no longer thinks in terms of life commitments to employees, nor expects the same from them. Automated teller machines and electronic fund transfer have begun to change job definitions at the clerical and teller levels. Tellers are now doing customer service marketing and are delegating the clerical tasks to technology. Clerks are being trained to become computer literate. As the use of technology becomes more extensive, fewer tellers and clerks are needed. Asked point-blank, a bank executive replied: "The only direction of change is fewer clericals."[9]

In a city like New York, with approximately 150,000 bank employees (1986 estimate) at all levels, these changes are adopted by banks in a fairly arbitrary fashion. An individual bank makes decisions regarding how it wants to use its staff. However, faced with stiff competition from financial services vendors, all confront the

need to reduce overhead. More technology and fewer workers seems to be the formula.

Looked at from a slightly different angle, the estimate that $700 billion per day passes through check clearinghouse and electronic fund transfer technologies in New York signals the widespread reliance on interbank technological networks. As a result of these technological transformations of banking, training programs are being used to attract workers who are familiar with computing and who have a high degree of visual activity. The latter involves numerical recognition skills, good eye-to-paper skills, and the ability to recognize relationships of scale, similarity, and difference. Bank training programs also continue to focus on self-presentation skills, which include the ability to articulate thoughts, visual presentation, and deportment. One bank in the New York area is reported to have screened almost 200,000 individuals to fill 10,000 positions. In a tight job market, the banks are in a position to look for employees with multiple skills.

One aspect of the banking industry reflects a shift away from job enhancement training to basic skill training. With more labor-intensive banking technologies, such as manual entry and tracking of funds, training was split. Professional training to acquire progressively more sophisticated information about industry trends and hierarchy-specific skills were available through postgraduate programs and seminars. For lower-echelon workers, the equivalent of an undergraduate curriculum in business and finance was available. With the widespread implementation of computer technologies, both training sectors have become more focused on technological know-how.

Ten years ago, entry-level positions in banking were filled by applicants likely to have a high school diploma or the equivalent. Today, these positions are awarded to college graduates or those with equivalent employment experience. So as automation is reducing jobs, placement criteria are becoming more sophisticated. The likely effect will be limited mobility in the industry.

The Resilience of the Work Ethic

Daniel Yankelovich and John Immerwarh have studied the American work ethic using survey methods that cut across a broad spectrum of the American population. The most striking conclusion to be

drawn from their study is that Americans want to work and to improve the quality of their work.

According to their study, *Putting the Work Ethic to Work* (1983), 63 percent of the American public believe that people do not work as hard as they did ten years ago. Sixty-nine percent reported that they feel workmanship is worse than it was ten years ago. While the results show lower assessments of work motivation, they also reveal the resilience of the work ethic: Seventy-six percent of the respondents reported a strong sense of dedication to their work. Furthermore, almost 60 percent reported that they have an "inner need to do the very best job I can, regardless of pay." What is surprising in these results is that, while 73 percent reported that work motivation is down, 60 percent would like to be more involved in efforts to get people to do their best on the job. More than 70 percent of those polled report that they would prefer to go on working even if they could live comfortably without working for the rest of their lives.[10]

Assuming that the Yankelovich and Immerwarh results are representative of American attitudes toward work, we seem to be heading for real trouble. The desire to work and the need for opportunities to demonstrate dedication and ability persist in a job market that offers far fewer highly challenging, well-rewarded jobs and many more routine, poorly rewarded ones. In such an environment, only workers with the appropriate technical or professional knowledge and credentials will get a crack at the limited number of desirable jobs and enjoy the rewards that are part of the work ethic. The lack of similar opportunities for the great majority of the workforce will produce widespread frustration, rage, and despair.

Changing Job Prospects

The November 1985 Department of Labor occupational employment projections for 1995 provide an indication of what the labor market in a high-tech society will look like.[11] They confirm the trend toward increases in the science and high-tech industries. Job losses in manufacturing and clerical occupations are an index of the fallout from technological unemployment.[12]

Among the thirty-seven occupations that will have the largest job growth, about one-fourth will require a college degree, which is about the present rate. Included in the list are:[13]

Occupation	Percent of total job growth
Cashier	3.6
Registered nurse	2.8
Janitor	2.8
Waiter and waitress	2.7
Accountant and auditor	2.9
Teacher, kindergarten and elementary	1.9
Computer programmer	1.5
General office clerk	1.5
Computer systems analyst, electronic data processing	1.3
Guard	1.2
Lawyer	1.1
Blue-collar supervisor	0.5

Although the need for workers in technologically oriented occupations is projected to grow, this increase does not account for most of the jobs projected to be added by 1995. According to Valerie A. Personick, an economist in the office of Economic Growth and Employment Projections: "Employment in high technology industries accounted for 6.1 percent of all wage and salary jobs in 1972, 6.4 percent in 1984, and is projected to represent 7.0 percent by 1995.[14] Put another way, by 1995 the number of high technology jobs will obviously not increase sufficiently to offset the jobs they will displace. In short, there won't be enough jobs and *89 percent of the total new jobs will not be in high-tech fields!*

Among the most rapidly declining occupations are these:[15]

Occupation	Percent decline in employment
Stenographer	40.3
Railroad brake, signal, and switch operator	26.4
Shoe and leather worker and repairer, precision	18.6
Sewing machine operator, garment	16.7
Farm worker	11.2

College and university faculty   10.6

Extruding and drawing machine operator, metal and
  plastic   9.1

The apparel and textile industries stand to lose about 350,000 jobs by 1995. Among the factors involved are technological improvements and foreign competition. The reason for the decline in college and university teachers is demographic: a result of the decline in enrollments. In part, this decline reflects the baby boom inflation of enrollments in the 1960s and 1970s.

When job-growth estimates and technological changes are looked at together according to job clusters, the following patterns are projected:

1. Technological advances, the graying of America, and the growing need for institutional alternatives for an aging population will account for job growth in health care occupations.

2. Clerical work will remain the largest major occupational grouping, accounting for an estimated 20.5 million workers in 1955. But it is an area of slow growth. Selvestri and Lukasiewicz note: "In addition to the direct impact that computerized office equipment will have on the clerical work force, the rate of employment growth of these workers is expected to be further slowed as more and more professionals and managers use desktop personal computers and executive workstations to do some of the work previously delegated to support staff."

3. Capital expenditures for equipment will probably provide increases in maintenance and repair work. The same pattern will be shown in automobile repair and maintenance work, even though there will be declining employment in automobile manufacturing.[16]

Do We Need A National Employment Policy?

Clearly, there are limitations in projections. These include the difficulty of estimating rates of technology implementation and the unavailability of accurate information on occupational trends. The Bureau of Labor Statistics researchers point out in their discussion

of data uses and limitations that factors such as replacement needs can often be a more likely source of employment than growth in an occupation. Their concluding statements focus on two issues that directly address the subject of a national employment policy: "Another consideration in interpreting the data on occupational demand is the availability or supply of workers trained or educated to enter an occupation. *Even with rapidly expanding job openings from either growth or replacement needs, jobseekers may have a difficult time finding a job because the supply of workers is expanding at an even faster pace.*" (Emphasis added)[17]

Taking into account the viability of the work ethic and the Bureau of Labor Statistics projections of shifts in employment, the American working population faces a difficult and contradictory situation. Work will be available, but not necessarily the work people are prepared for and, it appears, not at the wages they are conditioned to expect. It might be useful to consider here the discrepancies between the 72 percent of executive, administrative, and managerial employees who were displaced between 1981 and 1986 and found jobs and the 28 percent who didn't, or the 64 percent of blue-collar and factory workers now reemployed and the 36 percent still looking.[18] These numbers raise real questions regarding expectations of what is reasonable work and what level of compromises must be made by an individual to survive in changing job markets.

Unemployed skilled workers want jobs, but there is a growing mismatch of skills and occupational demand. (See the preceding tables.) The 1995 occupational projections indicate that what many of us have warned would be a result of automating workplaces will indeed happen. There will be a redistribution of jobs, with substantial reduction of the middle-level positions that provide sufficient income to support a middle-class lifestyle. Does this erosion of earnings, when combined with a seven to eight percent unemployment rate, constitute an unemployment problem that defies private, laissez-faire solutions?

The magnitude of the problem seems to dictate a redefinition of work that can be achieved only through the political process. One approach would involve changing work from an "obligation" to a "right," and redistributing jobs to provide sufficient employment for the entire eligible population.

When Senator Paul Simon of Illinois was asked about technological unemployment, his response included a historical observation that touches directly on the translation of work from an obligation and a means of meeting subsistence needs to something guaranteed to each citizen. He notes that many nations have been quicker to recognize their obligations to create jobs than the United States—the French realized it in 1848.[19]

Simon argues that there is a more sensible alternative to programs like welfare, which he believes subvert the right to work and alienate the majority of Americans who do work and resent those who don't. He proposes that people be paid for being productive, not nonproductive. His examples of how these people would work include:

Running day-care centers. Some parents could be personnel for these centers.

Teaching the functionally illiterate to read and write. Many out-of-work people who are highly literate could work under the supervision of a trained teacher. The millions of functionally illiterate people in this country are a substantial drag on the economy.

Providing transportation for senior citizens and the handicapped where adequate and appropriate transportation is not available.[20]

Simon calls for the establishment of local councils that combine representatives from labor, business, and the community to plan and oversee programs. While there are clearly political and administrative obstacles to the creation of such councils, they are not without precedent. For example, the Reagan administration's one domestic employment program to date, the Job Training and Partnership Act, has set in place the mechanism for involving local industry councils in training programs. Although organized labor has not been uniformly included, when it has become involved in council programs, the results have been encouraging. The Downriver Community Conference in East Lansing, Michigan, for example, has proven that such a coalition can work. Public–private partnerships have achieved notable successes, including the redesign of jobs and the retention of departing industries. But questions remain regarding the types of encouragement that local initiatives need.

What Simon proposes, however, goes deeper than a program recommendation. He is arguing that the right to work be acknowledged as a right of citizenship equivalent to freedom of speech and equality before the law. Secretary of Labor William Brock offered a much different viewpoint when asked if a federal policy is necessary to address the long-term effects of technological unemployment:

> I do not believe a federal policy is needed to address the long-term employment effects of technological changes. These effects, in fact, have historically been positive. Technological innovations and the related more advanced work methods, on the whole, generate more jobs than they displace.
>
> What is needed, and what we have, is a federal policy to assist workers who are displaced from long-term jobs as a result of technological change. Usually job displacement resulting from technology does not require government intervention. For those workers unable to adjust quickly to the new technology, the Job Training Partnership Act, administered by the Department of Labor, provides retraining and relocation assistance.
>
> [Note: In 1986 and 1987 the JTPA still remains the major conduit for federal interventions in the labor market.][21]

Brock's reply clearly shows the divide between the Reagan administration and earlier administrations of the past half century. The contrast between Simon's proposal and Brock's response also sharpens the debate over how extensively government should become involved in compensating for the differences between available jobs and available workers. It highlights the difference in political philosophy between Simon's belief in government's responsibility for the well-being of individual citizens and Reagan's faith in a self-correcting free market.

But whether Americans work because work is meaningful or because it provides them with the lifestyle they aspire to, or some combination of these and other reasons, they continue to work. Compared with West Germany, for example, the average American work week is thirteen percent longer. American workers typically receive fewer paid holidays than the Western Europeans or the Japanese (23 per year in the United States, compared with 30 in Japan, 34 in France, and 39 in West Germany).[22] Recent evaluations of the Job Training Partnership Act programs provide evidence that partici-

pants are serious in their motivation to find jobs. These evaluations also point out, however, that there are obstacles to finding jobs that prevent all applicants from being placed.

Inadequate job availability, the mismatch between required skills and qualifications of job seekers, and the failure to train workers in the skills needed are all factors that contribute to technological displacement. Doing enough to insure that those who want to work will have the right to do so demands a combined public and private sector initiative. A national employment policy can be built on the precedents established by successful programs like certain parts of JTPA and the earlier New Deal work programs. Political ideology must at some point come up against the statistical probability that work will need to be redistributed in the near future if our cultural values are to remain intact.

To surmount the unemployment caused by technological displacement, however, the major actors in the economy will have to cooperate. In order to evaluate and design an employment policy, federal policymakers, local policymakers, labor unions, and corporate leadership will need to work together. Some of the alternatives they must consider will threaten another core American value that has had increased emphasis since the late 1970s—the right to unimpeded action in one's own interest. But only with successful public–private cooperation can intervention to sustain work be translated into a guarantee of the general welfare. The Chrysler bailout in the early 1980s has created an awareness that even major companies cannot cope with changing economic conditions.

What argues most forcefully for a federal employment policy is the choice between support for public programs that will provide work and public programs that will provide welfare for those displaced by technology. Unemployment benefits are an exhaustible income-transfer response; they are short term and stopgap. Once they are exhausted, displaced workers and their families must draw on whatever resources they have. After that, the cycle leads to public assistance. Preferable alternatives can be created in publicly sponsored programs, such as the one outlined by Senator Simon. Also under consideration are various methods of redistributing work. Reducing the work day is one obvious solution; it would take into account the savings in time and effort that technology affords. Increasing the number of paid holidays is another option. Bringing the

U.S. workforce to parity with Western Europe and Japan in this regard is one form of intervention that should have a broad-based appeal for the majority of the labor force. Carefully planned sabbaticals that provide enough support and information to enable workers to improve skills are also an option, as is early retirement. Another idea is to encourage vocational education institutions to design programs that emphasize skills that will provide job security over a lifetime.

But the very attitudes that make work so important to Americans also make us reluctant to participate enthusiastically in redistributing it. Nor is industry, seeking international competitiveness through improved technology and corporate reorganization, eager to expand payrolls they have struggled to trim. Thus, a federal employment program could not be based, to any substantial degree, upon voluntary compliance. If we are to establish a right to work, then that right requires as much force in law as any other right.

Whether the U.S. government has the political will to establish such a right and whether our economy is capable of supporting it are questions for which there are no present answers. But weighed against a legislated right to work are the alternatives: not simply the higher cost of transfer payments, but widespread alienation, class conflict, and political instability.

8

The Social Crisis: The Polarization of America

THE 1980s will be remembered for several things. One, no doubt, will be the debate regarding the status of the middle class. We have seen endless analyses of occupational structure and earnings to confirm or deny the notion that the middle class is shrinking. Underlying these arguments is a concern that the path to upward mobility, the source of American economic vitality, is becoming less accessible. Thus, we have an economy that is tending to split society into the "little less than rich" and the "barely more than poor". As evidence are such present trends as the decline in industrial employment, the displacement of middle-level managers by communications-based technology, and the increased demand for unskilled service sector jobs. The future promises even more of the same: a continued erosion of the middle class that will radically change both the structure and the nature of our society.

Among those most concerned about the declining middle is Barry Bluestone, an economist and coauthor of the controversial *The Deindustrialization of America*.[1] In an *Atlantic Monthly* interview, he offered this diagnosis: "The pattern of wages in the old, mill-based economy looked just like a normal bell-curve. It had a few highly paid jobs at the top, a few low-wage jobs at the bottom and plenty of jobs in the middle. But in the new services economy, the middle is missing."[2] If the trend continues and accelerates, the long-term prediction is for a bulging underclass, a drastically shrunken middle class, and a somewhat larger upper class.

This redistribution is not solely the product of automation, but automation causes much of it, and management uses automation as a scapegoat for more. The entrance into the labor force of the baby boom generation is also a factor. It has provided an oversupply of candidates for middle-level positions and created a buyer's market in which wages are inevitably depressed. Another factor is economic change caused by international trade. Because of the high cost of labor in the United States, a number of jobs have moved abroad. While these were blue-collar jobs, they had paid wages sufficient to support middle-class lifestyles. Meanwhile, automation is taking its toll in shifts from labor-intensive to machine-intensive production, and in office automation, which is trimming middle management. And the growth of the service sector of the economy accelerates the process. The now-discernible trend shows that displaced workers who find new jobs frequently earn less and receive fewer benefits.

A Stanford University study (1984) on technology and the design of jobs in the insurance industry uses hard data and leaves little room for debate:

> Office automation has wiped out thousands of jobs for low skilled clerical workers, created new jobs for skilled clerical workers, and eliminated many professional jobs that comprised the middle of the occupational pyramid and that used to constitute the rungs of a career ladder by which clerical workers could climb up into more highly skilled professional jobs. . . . As a result of the automation of underwriting and claim estimating for standardized insurance products, career ladders from skilled clerical to insurance professional positions have been eliminated.[3]

In high-tech industries, which account for only a small proportion of the labor force, some interesting patterns are emerging that support the bipolarization argument. It was once thought that legions of jobs would be created for engineers and technicians, who would take care of the new high-technology equipment—programmers, service technicians, and maintenance engineers, for example. But the projected difference between the need for janitors in 1995 (500,000) and for engineers (115,000) shows how limited the opportunities for middle-class and professional jobs will be.[4]

The computer industry, the greatest area of high-tech expansion, also reveals less promise than many had expected. Part of this has

to do with the technology itself. With chips, much of what would normally need to be repaired or maintained is miniaturized on an interchangeable piece of silicon. The technology is being perfected to the point at which maintenance and upkeep will become minimal. And the industry is developing a clear bipolar pattern: there are jobs for highly skilled professionals and technicians, low-skilled assembly work jobs that require little experience, and little in between. Furthermore, if the promises of artificial intelligence are even modestly realized, jobs affected by this field are likely to contract rapidly.

Another field in which it appears that expansion will not match expectations is clerical work. In the early 1980s it was considered a boom occupation, but Bureau of Labor Statistics estimates now indicate that demand is already starting to recede. A study called "Crisis and Computerization: The Uncertain Future of Clerical Work in the Great Lakes States" indicates the same pattern. Jobs have contracted by close to fifteen percent.[5] This case is stated most strongly in the draft summary of a study called "Impact of Office Automation on Office Workers" (1984) conducted at the Georgia Institute of Technology:

> Under the most conservative assumptions, clerical employment in 2000 is reduced below the 1980 figures in both industries [banking and insurance]. At minimum, we foresee *absolute* reductions in clerical employment of 22% in insurance and 10% in banking by 2000. *Beginning about 1990 workload estimates continue to increase while clerical employment is forecast to decline, a situation different from any faced in recent decades.*[6]

Computer maintenance engineers, when asked about the impact of the chip technologies on their jobs, have some interesting responses. They are at first generally enthusiastic about the capabilities of the technology. Then, as we discuss what is changing in their work, what they say betrays their insecurities: the realization that they are not only replaceable, but disposable. "Formerly, you had to know everything about the technology, now it's different. We rely on another unit to rebuild circuit boards—then it's a question of seeing if they did their job right." Inserting a substitute board and recalibrating a computer or local area network is still a skilled job, but they clearly wonder how long it will remain one.

Downshifting and Deskilling:
Less and Less on Every Level

The engineer's assessment of his job situation touches on one of the key concerns of the declining middle thesis. Complementary to the adoption of high technology is a process of "deskilling," or a "downshifting" of skills that had formerly commanded higher wages. Secretarial skills such as calendar management and file management, for example, are being taken over by the computer. As a result, the secretary's job more closely resembles that of the clerk. Assembly line or "low-tech" word-processing shops where workers are often part-time and have no benefits are another example. These workers tend to have minimal involvement in the definition or evaluation of their work. In the case of the computer maintenance engineer, deskilling can be seen in his dependence on the unit that rebuilt the circuit board.

Such deskilling, which involves the loss of valuable knowledge and the exercise of control over work, reduces the quality of work life as well as wages. Western society has emphasized work as the path to self-esteem in addition to economic well-being. Numerous studies conducted in Europe as well as the United States indicate that the downshifting of jobs to more menial and routine activities leaves workers with reduced expectations and a diminished sense of self-worth. The loss of challenge and mastery, which has been ceded to intelligent technology, creates morale problems. When jobs can be reconfigured and parts of them performed by computers, the usual result is higher profit ratios and, if the marketplace so directs, lower prices to the consumer. Questions about the role of work in that same consumer's life and the ultimate effect altering that role will have on the market are seldom analyzed.

Jim Knowles, who works as a systems analyst at *The New York Times*, describes some of the automation that has taken place there. It can be taken as a model for the declining middle class thesis: "Basically key punch jobs have been relegated to clerks within any given office, thereby eliminating the need for key punch operators. At the same time, systems analysts have been "red-lined" into management positions. The effect is not only to reduce the number of jobs but also to create clear division between management and labor."[7] Knowles, who has retained his status as labor and is a shop

steward, feels that this is an irreversible trend. He managed to keep his position a union one, but he is unique. The next step at the *Times*, he feels, will be the progressive farming out of clerical work. This pattern was set by the home delivery and subscriptions services, both now handled by outside firms.

The Return of the Cottage Industry: Farming Out

An even more direct assault on the employees' view of work in an organization comes from farming out assignments to freelancers. Recent approaches to management that advocate team playing and employee participation will work only when the quality of work life is taken seriously. Working in a "winner" or an excellent organization and integration into the corporate family are important to employees; they provide the fringe benefits of sociability and a sense of commitment. When management farms out work to freelancers at home-based computer terminals, these fringes are eliminated. At the same time, the split between employees accentuates the differences between those who belong to the organization and those who don't. The obvious contradiction between the ideologies of worker involvement and the revival of off-site piecework methods of data processing—now practiced in the insurance, publishing, airline, and banking industries—are but two examples of the gap between management rhetoric and reality.

In some industries, "farming out" is done at the workplace: freelancers are hired to work in the company facilities, much like temps. They are not considered employees and are paid an hourly or piecework rate as independent contractors. Companies save a great deal of money on fringe benefits (no paid vacations, sick days, or insurance), overhead (office space for temps does not have to be attractive or comfortable), and investments in personnel (training and seminars). But freelancers, like temps, have no stake in the company nor any incentive to contribute their talents or ideas to the success of the enterprise.

Farming out work is a complicated aspect of polarization. For professionals and middle-level managers, the ability to pace themselves and take work home enhances job autonomy. The type of work that these people do has never been suited to an accounting system that pairs output with time spent. For such workers, the

convenience high technology affords is an asset. The fact that the work they do on computerized systems is integral to the goals of the organization further enhances their identification with their employer.

At the other end of the scale, however, is the farming out of piecework, such as processing medical reimbursement claims or tallying airline ticket information. Clerical tasks that involve minimal discretionary judgment require little identification with the employer except a concern for accuracy and efficiency. Computer monitoring provides measures for assessing both. These jobs are the most subject to deskilling; they are being redefined to allow for the least possible human error.[8] The process is parallel to the automation found in numerically controlled machines, in which the ideal condition is an autonomous machine that can monitor itself.

Management rhetoric of participation creates problems when workers feel psychologically displaced from their work and the workplace. For those who lose their jobs, the message is clear; but for those whose jobs are deskilled, the effects are complicated. A management policy of participation would lead workers to think that they have a stake in decision making, while their job description says otherwise. Anger and disappointment at losing control over work are manifestations of loss of pride and job satisfaction, which are tied to deskilling.

These emerging patterns—which have to do with what workers expect their jobs to be, as well as with the job definition in a personnel file—are found in white-collar as well as blue-collar occupations. They underline the value that work plays as a source of self-worth. The middle manager whose job is phased out as part of a conversion to "lean and mean" management is now potentially brother and ally to the automobile welder who oversees the robot that replaced three men. Despite promises of a rewarding leisure society, American values still are linked very much to the assumption that "work makes the woman or man."

Relocation—And Dislocation

Another aspect of the declining middle debate involves the relocation of industries from the rust belt, first to the sunbelt and now to the middle Atlantic regions of the country. Tennessee, for example, is clearly a boom state in the 1980s. One good reason is that it has

actively sought industry; a second is that its workforce is relatively unorganized. Energy is also cheap and abundant.

Tennessee's good fortune is a loss for Michigan and the rest of the industrial North and Midwest. There, departing industries leave behind some difficult issues for policymakers, including the extent to which workforce relocation should be encouraged, the design of programs and incentives to create new jobs to replace those lost, and the provision of training programs to supply workers for these jobs.

Workforce relocation is an emotional as well as a strategic part of this policy equation. Mobility, or lack of it, involves community, investments in property, and the importance of friends and family as a support system. To pack up and move requires not only money, but sufficient resilience to dispense with the ties that come over time from living in a community. In this respect, younger members of the workforce clearly have an advantage, since they are at the stage of life when moving around to find the right job is much easier. They simply don't have the obligations that come with being settled in a career and established in a community.[9] To the extent that they are free from the responsibilities of children and dependents, younger workers are more mobile economically as well.

The deindustrializing areas of the country provide some provocative information on this issue. In the smokestack industries, it is not uncommon for one generation of the same family to follow the last one into the plant. Historically, coal mining, steel, and the automotive industry, as the major employers in these areas, absorbed most of the available blue-collar workforce. After a high school education, most skills were learned on the job. Work in the same industry was the normal pattern for many families. Once people settled into these well-paying, secure jobs, they tended to buy property and raise families. Wages were generally high enough to allow a middle-class lifestyle.

When these industries began contracting, many of these strong ties transformed into obligations that became difficult to support. Mortgages that were manageable at a full salary become impossible with unemployment benefits. Property sometimes had to be sold at a loss, if it could be sold at all. Investments made during strong earning years diminished as the local economy went sour. In certain areas around Detroit, for example, houses purchased for $50,000 in the early 1970s are now selling in the mid-1980s for $25,000.[10]

Other Costs:
The Resurgence of Discrimination

From the sociological point of view, a decision to relocate is complicated by the rewards and responsibilities that come from being a member of a community. Also complicating the decision to move are ties to relatives; participation in community groups like local sports associations, volunteer fire departments, or Lions Clubs; children's involvement in their schools and with their friends. Such factors often keep workers in a town or region when it is clear that they will never regain the jobs they once held.

Migration to available jobs is one solution. The creation of new jobs is another. Economic trends indicate that if new jobs come, they will be in the service economy, where lower skills, lower pay, and fewer benefits can be expected. Either downshifting of lifestyle to match income, or dual-earning families, or both will result. With more than fifty percent of adult women now in the labor force, reliance on double incomes has already become a means of maintaining lifestyle. According to Bureau of Labor Statistics data, this too has its depressing side. Women and youth are the most likely to have part-time and low-paying jobs. They are often the first to be laid off in any industry that uses a "last hired, first fired" seniority system. They are most likely to be in the positions that are quota-driven, service-oriented, and demand intense effort.[11]

Minority group members, especially black males, have found employment in similar types of jobs. Furthermore, a 1986 Office of Technology Assessment report on displaced workers shows that blacks are at a comparative disadvantage in finding substitute jobs. One study shows that, among all workers who had held jobs for three or more years, sixty percent of those displaced found new jobs, but only forty-two percent of black workers who had held jobs for at least three years found employment.[12]

The vulnerability of women and minority groups means that bipolarization will strike them hardest. Gender and race discrimination is reemerging as a threat, particularly in the information industries, where white males occupy most senior positions. Another, perhaps ironic, example of perceived discrimination has been the movement of many corporate clerical positions from inner-city areas to the suburbs. The subtle arguments behind this change include

the more industrious work habits of suburban housewives, their compliance, and their passive role in bargaining. In contrast to their inner-city (usually black) counterparts, these women are more likely to be part of dual-income households and able to manage better on part-time wages.

The Other Side of the Argument

Not everyone agrees that bipolarization is an irreversible trend. Neal Rosenthal, in a provocative piece called "The Shrinking Middle Class: Myth or Reality?" (1985) presents evidence that, although there have been reconfigurations within occupational and earnings distributions, "my analysis of occupational trends for the 1973–82 period shows that the tendency toward bipolarization, if it did exist, seems to have been reversed by the mid-1970s."[13]

In a review of the bipolarization debate, Rosenthal points out that our economic structure is very complex. Factors other than the decline of employment in smokestack industries, the rapid growth of high-tech industries, the large numbers of job openings, numerical growth in low-paying occupations, and the shifting industrial structure from goods-producing to service-producing industries must be taken into account. These factors include Bureau of Labor Statistics data indicating that nearly all production workers in high-tech industries earn hourly wages higher than the average for all manufacturing jobs. In 1982, the total average for high-tech industries was $8.50 per hour, compared with $7.68 per hour for all private non-agricultural establishments. Other evidence is the projection that the twenty occupations slated for fastest growth in the period 1982–1995 are distributed predominantly in the top and middle third of earnings categories, even though most are service-related. But, based on our own analysis of the statistics, particularly over the last four or five years, Rosenthal's thesis does not hold true for the late 1980s and 1990s—and neither do the earlier projections for this period.

When we asked Secretary of Labor Brock, he drew a distinction between the appearance of a dual labor market phenomenon and the trends as indicated in the Bureau of Labor Statistics statistical analysis. We quote his answer at length because it points out the complexity of relying on projections:

At first glance it seems as if the United States is losing manufacturing jobs that support the middle class and creating only lower-
paying, less skilled service sector jobs. Our economy seems to be
polarizing into primary and secondary labor markets of the "haves
and have-nots." A recent study by the BLS in 1983, however,
shows that the tendency toward bipolarization, if it did exist, seems
to have been reversed since the mid-1970s. Some trends in the
industrial and occupational structure of employment could cause
a degree of bipolarization. However, a multitude of factors have
had an effect on the occupational structure of our economy. The
BLS analyses of available data indicate that the combined effect of
all factors apparently had not caused bipolarization over the 1973–
82 period. Also, given BLS projections of employment by occupation, bipolarization is not likely to occur between 1982 and 1985.[14]

No matter how many factors are analyzed, to the workers involved no factor means more than buying power. And, measured in
buying power, the current trend is clearly downward. If we look at
families with incomes of between $15,000 and $35,000 per year, we
find that the proportion dropped from fifty-one percent of total
families in 1973 to forty-four percent in 1982 (the calculation was
made in constant 1982 pretax dollars). At the same time, the extremes
at both ends grew. At the individual family level, an example from
the steel industry is instructive: A laidoff steelworker who finds a
job in the electronics components industry is likely to take a pay cut
from $13.50 to $7.50 an hour (1983 dollars). Add to this lost seniority
rights and the probability that benefit packages will be reduced, and
we are talking of a substantial drop in standard of living.[15]

The Conference Board projects that by 1995 about one-third of
personal income will be earned by households with total incomes of
$50,000 or more (in constant 1982 dollars).[16] The trend toward the
expansion of professional and technical middle-class occupations par-
allels the erosion of the blue-collar middle-class jobs in the manufacturing sector. In the decade between 1973 and 1983, nine out of
ten new jobs were created in the service and trade sectors of the
economy. More than half of this growth occurred in four industries,
which *Fortune* writer Bruce Steinberg calls the "new tetrarchs of the
labor market": health, business services, finance, and food services.
The blue-collar middle-class worker with skills not easily transferred
to professional/technical levels will find jobs in these generally lower-

paying nonprofessional, nontechnical sectors of the economy—if at all.[17]

This translates into diminished buying power for households that formerly consumed at middle-class levels. In terms of buying power, quality of life is not the only issue. Such changes pose problems for manufacturers of durable goods. Faced with the challenge of maintaining levels of consumption in the face of the possible loss of a mass market, manufacturers of automobiles and refrigerators are confronted with a dilemma. They are directing the industries that provide fewer and fewer jobs for the blue-collar middle class. Their policies are directly responsible for some of the shifts in income. The result is a powerful Catch-22 that to date has acted only to fuel competition between producers.

Lifestyle and consumption are powerful correlates of class status. They provide evidence to a community of where someone fits. Cutting back on durable commodities is a form of retrenchment, and residents of depressed areas all recognize it as such. Makers and marketers of disposable goods have been canny in their appreciation of this and have developed stratified marketing strategies to adjust to the shrinking of a middle-level consuming group.

The trends toward polarization of the American class structure are real. Too many powerful empirical indicators argue against dismissing it. For example, on September 22, 1986, the *Wall Street Journal* ran this front-page article: "Growing Gap: U.S. Rich and Poor Increase in Numbers: Middle Loses Ground." The reporter, David Wessel, surveyed the various economic interpretations of a declining middle and offered the following data: "Last year, the top fifth of American families—those earning more than $48,000—got 43% of all family income, a post war high. The bottom fifth—earning less than $13,200—got 4.7%, the least in 25 years. Families earning $15,000 to $35,000 a year, adjusted for inflation, fell to 39% of the total last year from 46% in 1970."[18]

Before the bifurcation of the nation proceeds much further, we need to recognize that the nature of our society depends upon its structure and an economy which cannot preserve that structure does not serve the nation. If the price of becoming internationally competitive is to unravel the fabric of our society and erode our domestic market, then protectionism may not be so horrendous an alternative. We must bear in mind that economic goals must be formulated to

assure the well-being of the nation—the society—and not its corporations. There's something dreadfully wrong when what we perceive as good for the economy is bad for most Americans.

Conclusion

From its beginnings, the computer has been surrounded by fantasies, wishes, projections—and the wariness that comes from knowing what can happen when machines go out of control. The computer promised to be a machine that could think, one that would not only solve problems, but create a few. As computers have become a more regular and accepted part of daily life and the workplace, most of these projections have been shown to be inflated. With the 1986 *Challenger* disaster and with the increasing visibility of technological unemployment, Americans have been painfully reminded that technology certainly is not perfect: seldom does it live up to its promises. But once we can see it as a tool and as a means of performing work, we can begin to judge the computer realistically. The realities of the workplace and the goals of the workforce must include a responsible analysis of technology.

The argument of this book is that until leaders in industry, government, and labor begin to conduct such analyses and to understand the uses, real possibilities, and limitations of technology, it will continue to create as many problems as it solves. High-technology has the potential for dramatic labor savings and is therefore a factor in capital expense and profit equations. Translated into the human terms of unemployment and relative deprivation, it must be considered in the total context of both the benefits it provides to society and the penalties it imposes.

Technological unemployment to date has been perceived as being restricted to areas that are either deindustrializing or revamping to service and non-labor-intensive high-technology employment. But

its effects, while they may still be relatively limited, are profound. The minimally funded Title III provisions of the Job Training Partnership Act are evidence, as are the inquiries conducted by numerous economists and social scientists.

But beyond the specific effects of technological unemployment are the social benefit issues being raised. They pose basic questions about economic goals and the price that we, as a society, can be expected to pay for new technologies that may serve many Americans more poorly than the old.

Among the questions to be resolved is how much responsibility the government bears for insuring that citizens have the right to work. The projected impact of high technology on jobs that have traditionally provided middle-class wages—either blue- or white-collar—is that there will be fewer of them. While some will be replaced by new high-technology professional and highly skilled jobs, many more will be replaced by service jobs that provide lower wages, fewer benefits, less security, and require minimal skills. We have argued that this will result in the decline of the middle class and possibly in the restructuring of our traditional upwardly mobile society. Should the latter occur, our democratic institutions might well be at risk.

In calling attention to these possibilities, our goal is to foster a long- and short-term debate on the technological transformation of the workplace. We are also convinced that the "up by the bootstraps" mentality of corporate and government leadership is at best short-sighted and at worst inhumane. Americans have come to expect that work will be available and that it will be rewarding. Historically, this optimism has fueled much of our national growth. From both an economic and a social perspective, it would be counterproductive to jeopardize these expectations—especially when the U.S. economy needs to reassert a competitive presence in international markets.

Whether or not Americans as a society can agree on provisions for a "right to work" comparable to those guaranteed in the industrially developed countries of Western Europe and in Japan is a key factor in properly confronting the impact of high technology on the workplace. To the extent that we retain a laissez-faire attitude and assume that the market will correct itself, questions of the quality of work, a fair level of social benefits, and the regulation of working

conditions remain marginal. If, however, a commitment is made to provide some type of right to work legislation, then programs must be designed and implemented on a large scale.

Such a commitment involves political decisions that can only be made with widespread public support. But most Americans appear quite content to go about their lives without questioning what the changing composition of the workforce means. The news media does not dwell on changes in labor patterns, while it does focus on how high technology will improve our lives. Computerized cars remind us of maintenance requirements and refrigerators tell us when the door has been left open. Generally ignored is the question of who will get to enjoy all the new consumer commodities. The projected polarization of the labor force leads us to believe that only those at the upper end of the income scale will have the means to enjoy such benefits.

Society's awareness or unawareness of questions like this is in many ways a litmus test of how well we are able to identify political and moral issues. When we look at the future as the futurists do, we tend to see a new frontier. This is an attractive vision. It is closely tied to the image of our historical frontier and represents a continuation of Manifest Destiny, a new opportunity to conquer uncharted territories.

But with the exception of perhaps the ten to twenty percent of jobs projected for 1995 as high-technology-related and offering mobility, this new frontier is in fact the one we have lived with since the introduction of automation in the post-World War II period. While there was then concern about large-scale unemployment, now we tend to read about job displacement. But always we are sure that there is an easy solution, and this complacency is the most dangerous aspect of our preoccupation with new frontiers and the marvels of the new technologies.

As social critics like Roszak, Shaiken, and many others have argued, the time has come for high technology to be realistically positioned.[1] If the computer does change the nature, quality, and status of work in American society, then its impacts must be charted and anticipated. The need to introduce some level of predictability into broad social changes might, in actuality, be what makes the futurists' books such best sellers. To use Toffler's metaphor, people would rather ride the wave than be crushed by it.

Our examination leads us to argue that the United States must not only mitigate the effects of social changes resulting from the adoption of new technology, but must control them. We first must provide a safety net for those whose jobs are displaced, down-scaled, or abolished. Traditional training and retraining programs can only be part of this net; the effort should also involve mainstream educational institutions that bear the responsibility for preparing the workforce. We need to recognize that what is really called for is life-long education to maintain and improve job skills. Educational institutions must be encouraged and funded to develop programs that upgrade skill levels of the workforce while creating new employment options.

But if we are really to control the social impact of the new technologies, then we must seriously consider a societal commitment to full employment that goes beyond policy to law. A right to work is no more than the minimum price society should be demanding. Beyond that basic right, we must seek ways to halt the erosion of the middle class and preserve the structure that has given our society its strength, supported our democratic institutions, and allowed us to avoid the class conflicts and political instabilities that plague most other nations.

Afterword

D URING the preparation of *Dreams Betrayed*, and after Secretary of Labor William E. Brock had responded to our queries, the Department of Labor completed a report on the workforce in the year 2000. Although the Labor Department researchers anticipate many of the same changes both in the workforce and the job market that we do, there are some substantial differences between what we believe the impact of these changes will be and what the department predicts.

The report, "Work Force 2000," predicts, for example, continued regional shifts in employment, with growth greatest in the South and West and least in the Midwest. Jobs will continue to shift from the production to the service sector of the economy, with ninety percent of new jobs in services and only eight percent in manufacturing. The report sees a minimal expansion of the workforce, which is expected to grow at no more than one percent annually by the year 2000. As fewer younger men and women enter the job market, the average age of workers will rise, and the department predicts that between now and the end of the century, "most labor force growth will come from groups in the population that have traditionally been underutilized. Women, minorities and immigrants will account for more than 80 percent of net additions."

The department's scenario includes, as does ours, "rapid turnover and changes of industries" that will require workers to adjust to new jobs more often and more rapidly. According to the workforce report, some workers will be "changing jobs five or six times during their worklives."

But the Labor Department's assumption that "new and existing jobs will require higher levels of analytic skills" does not square with the kinds of changes in employment anticipated both by the department and by us. While a great many skilled jobs in production will indeed be lost in the years ahead, it is unlikely that more than six percent of the workforce will be employed in true high-tech positions. The department sees this illusory demand for more highly skilled workers as "a window of opportunity" that would solve some long-standing problems by providing "higher wages and more hiring and training opportunities for less well-prepared workers" (i.e., "minorities, the handicapped, immigrants, and women").

As a result, "Work Force 2000" proposes that there is a good chance for "significantly narrowing the occupational earnings gaps that have separated specific groups of the population from the mainstream." The Labor Department sees the likelihood of society pulling closer together and of the expansion of the middle class.

This stance is in line with the official optimism of the Reagan years. Unfortunately, however, we foresee job market changes having quite the opposite effect. We believe the loss of skilled jobs, the shift of more workers from production to service, and the limited employment opportunities provided by the new technology will result in more worker downgrading than upgrading. Instead of an opportunity for the middle class to expand, we argue in *Dreams Betrayed* for the likelihood of middle-class erosion, as the number of both upper-class and lower-class Americans increases.

Notes

Introduction

1. Paul O. Flaim and Ellen Siligal, "Displaced Workers of 1979–83: How Well Have They Fared?" *Monthly Labor Review* (June 19, 1985).
2. *Making America Work Again: Jobs, Small Business and the International Challenge* (Washington, D.C.: National Commission on Jobs and Small Businesses, 1987), p. 13.
3. Ibid., p. 7.
4. Mitchel Moss, "The Telecommunications Infrastructure in the City of New York," New York University, submitted to the Division of Policy Analysis, Office of Economic Development, City of New York (1985), p. 1.
5. Schultze's comment was in response to a question posed by Roger Williams, 1984. The Brock letter was in response to a question on technological displacement, November 5, 1985. All research done for Carlton Rochell.
6. From Martin Mayer, *The Bankers* (New York: Weybright and Talley, 1974).
7. Robert Kuttner, "The Declining Middle," *Atlantic Monthly* 252 (July 1983), pp. 60–72.
8. Bureau of Labor Statistics projections from *Monthly Labor Review* 108, no. 11, (November 1985).
9. Diana Rouse, research director, "Crisis and Computerization: The Uncertain Future of Clerical Work in the Great Lakes States." Final report submitted to the Joyce Foundation by the Working Women Education Fund, May 15, 1985.

Chapter 1

1. The succeeding history of computer automation is drawn from the following sources: Stan Angarten, *Bit by Bit: An Illustrated History of Computers* (New York: Ticknor and Fields, 1984); Roger Draper, "The Golden Arm," *New York Review of Books* (October 24, 1985), pp. 46–52; Christopher Riche Evans, *The Making of the Micro: A History of the Computer* (New York: Van Nostrand Reinhold, 1981); Raphael Kaplinsky, *Automation: The Technology and Society* (Harlow, England:

Longman, 1984); Lenny Siegel and John Markoff, *The High Cost of High Tech* (New York: Harper and Row, 1985).

2. Philip Elmer De Witt, "A Sleek, Superpowered Machine," *Time Magazine* 125, 4 (June 17, 1985) p. 53; and Lucia Solorzano, "Computers on Campus: The Good News, Bad News," *U.S. News and World Report* 99, 15 (October 7, 1985), pp. 60–61.

3. National Computing Centre, *Impact of Microcomputers on British Business* (Manchester, England, 1979). Quoted in Raphael Kaplinsky, op. cit. p. 91.

4. Roger Draper, op. cit., p. 47.

5. Jack Hollingum, *The Engineer* on Caterpillar Tractors, quoted in Raphael Kaplinsky, op. cit., p. 90.

6. Howard Banks, editorial, *Forbes* (October 25, 1985), p. 32.

Chapter 2

1. For the impact of G.I. Bill, see James Fallows, "The Case Against Credentialism," *Atlantic Monthly* (December 1985), pp. 49–67.

2. This discussion of American affluence is based partly on John Kenneth Galbraith's book *The Affluent Society* (Boston: Houghton Mifflin, 1958).

3. David Reisman, "Leisure and Work in Post-Industrial Society," 1958, in *Abundance for What? and Other Essays* (Garden City, N.Y.: Doubleday, 1964), p. 172.

4. Study cited in ibid., pp. 169–170.

5. Daniel Bell, *The Coming of Post-Industrial Society: A Venture in Social Forecasting* (New York: Basic Books, 1973).

6. Ibid., pp. 14–45.

7. The phrase "time is money" recurs throughout the literature on modern industrialism. Many believe that it was coined by Benjamin Franklin.

8. Wassily Leontief, "The Long-Term Impact of Technology on Employment and Unemployment." Speech delivered at the National Academy of Engineering Symposium, June 30, 1983.

9. Andrew H. Malcolm, "What Five Families Did after Losing the Farm," *The New York Times* (February 4, 1987), pp. A1, A18.

10. Congressional Research Service, *The Computer Revolution and the U.S. Labor Force.* For the use of the Subcommittee on Oversight and Investigations of the Committee on Energy and Commerce, U.S. House of Representatives. (Washington, D.C.: U.S. Government Printing Office, 1985).

11. Ibid., p. 5.

12. Ibid., pp. 36–37.

13. Ibid., p. 38.

14. The notion of the "laboring and dangerous classes" comes from Louis Chevalier, *Classes Laborienses et Classes Dangereuses* (Paris: Librarie Plon, 1958).

15. David Reisman, op. cit., p. 166.

16. Roper and the American Council of Life Insurance, 1983.

17. U.S. Department of Health, Education and Welfare, *Work in America* (Washington, D.C.: U.S. Government Printing Office, 1972), p. 59.

Chapter 3

1. Alvin Toffler, *The Third Wave* (New York: William Morrow, 1980), ch. 1, "Super Struggle," pp. 25–34.
2. Ibid., p. 32.
3. Ibid., pp. 25–34.
4. Ibid., ch. 16, "The Electronic Cottage," pp. 210–223; and ch. 17, "Families of the Future," pp. 224–242. The catalog of options is on p. 241.
5. Ibid., p. 104.
6. John Naisbitt, *Megatrends: Ten New Directions Transforming Our Lives* (New York: Warner Books, 1982, 1984).
7. Ibid., p. xxxii.
8. Ibid., p. 36.
9. Ibid., p. 36.
10. Ibid., p. 43.
11. Ibid., p. 39.
12. Herman Kahn, *The Coming Boom: Economic, Political and Social* (New York: Simon and Schuster, 1982).
13. Ibid., p. 64.
14. Ibid., ch. 3, "The Basic Context for Revitalization," pp. 50–64.
15. Ibid., pp. 51–53.
16. Ibid., p. 52.
17. Isabel V. Sawhill, "Reaganomics in Retrospect," in *Perspectives on the Reagan Years*, ed. John L. Palmer (Washington, D.C.: The Urban Institute Press, 1986), p. 113.
18. Herman Kahn, op. cit., p. 71.
19. Ibid., pp.72–73.
20. Ibid., pp.72–85.
21. Robert J. Samuelson, "Our Computerized Society," *Newsweek* (September 9, 1985), p. 73.
22. A.R. Martin quoted by Robert Theobald, "Cybernetics and the Problems of Social Organization," in *The Social Impact of Cybernetics*, ed. C.R. Dechert (New York: Simon and Schuster, 1966), pp. 55–56.
23. Stanley Aronowitz, Dana Fenton, and Bill Kornblum, "Preliminary Findings and Recommendations Regarding the Possible Effects of Computer-Aided Design (CAD) on Members of Local 375 AFSME, District Council 37" (New York: Institute on Labor and Community, May 18, 1984), p. 3.
24. Simon Nora and Alain Minc, *The Computerization of Society: A Report to the President of France* (Cambridge, Mass.: MIT Press, 1980).
25. Pierre Bourdien, "Bourdien: A quand un lycée Bernard Tapiez," *Liberation* (December 4, 1986), p. 4.
26. Simon Nora and Alain Minc, op. cit., p. 129.
27. Ibid., p. 131.
28. Clark Kerr, *The Future of Industrial Societies: Covergence or Continuing Diversity* (Cambridge, Mass.: Harvard University Press, 1983), p. 126.

Chapter 4

1. Jerry Flint, "We're Going To Do Just Fine," *Forbes* (October 21, 1985), pp. 37–40.
2. Quoted in John Nielsen, "Management Layoffs Won't Quit," *Fortune* (October 28, 1985, pp. 46–50.
3. Peter Nulty, "Pushed Out at 45—Now What?" *Fortune* 115, no. 5 (March 2, 1987), p. 26.
4. International Labour Organization, *World Labor Report 1: Employment, Incomes, Social Protection, New Information Technology* (Geneva, Switzerland: International Labour Organization, 1984).
5. Harley Shaiken, *Work Transformed: Automation and Labor in the Computer Age* (New York: Holt, Rinehart and Winston, 1984).
6. John Naisbitt and Patricia Aburdene, *Re-Inventing the Corporation: Transforming Your Job and Your Company for the New Information Society (New York: Warner Books, 1985).*
7. *Ibid., the argument runs throughout the book.*
8. *Robert Howard, Brave New Workplace* (New York: Viking, 1985), p. 96ff.
9. Ibid., p. 71ff.
10. Stanley Hyman, quoted in John Nielsen, op. cit., p. 47.
11. The discussion of farming out by Blue Cross is from Robert Kuttner, "The Declining Middle," *Atlantic Monthly* 252 (July 1983) pp. 60–72.
12. Fritz Hauser quoted in Denis Chamot, ed. "Technological Change and Unions: An International Perspective," AFL-CIO Department of Professional Employees, Publication # 82-2, pp. 13–16.
13. Kathleen K. Weigner, "John Young's New Jogging Shoes," *Forbes* 136, 4 (November 4, 1985), p. 43.
14. For a concise overview of trends in clerical work, see the U.S. Department of Labor, *Women and Office Automation: Issues for the Decade Ahead* (Washington, D.C.: U.S. Government Printing Office, 1985), p. 22.

Chapter 5

1. Quoted in "Troubles of U.S. Labor Unions Eased in 1986," *The New York Times* (February 15, 1987), p. 31.
2. Statistics on union membership come from "The Changing Situation of Workers and Their Unions: A Report on the Future of Work," prepared by the AFL-CIO Committee on the Evolution of Work (Washington, D.C.: AFL-CIO, August 1985), p. 5.
3. Ibid.
4. Letter, December 20, 1985, p. 4.
5. Steven Greenhouse, "Painful Moves in Steel Go On," *The New York Times* (December 29, 1985), p. D2.
6. Quoted in Kenneth B. Noble, "Labor Department Data Show a Revival of Major Strikes," *The New York Times* (September 21, 1986), p. 31.
7. A summary taken from Kenneth B. Noble, "Steel Union Locals Back USX Pact," *The New York Times* (January 19, 1987), pp. D1, D3; and Leslie Wayne, "Implications of USX Pact," *The New York Times* (January 22, 1987), p. D2.

8. Peter Unterweger, "Work, Automation and the Economy," paper presented at the American Association for the Advancement of Science, annual meeting, Detroit, Michigan (May 27, 1983), p. 2.
9. Anne B. Fisher, "Ford Is Back on Track," *Fortune* 14 (December 23, 1985), pp. 18–22; Pestillo quote, p. 22.
10. Resolution adopted by the 27th UAW Constitutional Convention, May 15–20, 1983, Dallas, Texas. Copy obtained from UAW Research Department.
11. Ibid.
12. AFL-CIO Committee on the Evolution of Work, op. cit. p. 5.
13. Ibid., p. 6.
14. Ibid., pp. 12–13.
15. Ibid., p. 13.
16. Ibid., pp. 10–11.
17. Michael Brody, "Meet Today's Young American Worker," *Fortune* (November 11, 1985), p. 92.
18. Cited in Charles L. Howe, "Big Labor and Big Blue," *Datamation* (January 1, 1986), p. 30.
19. Ibid., p. 32.
20. Discussion with Rochell on January 27, 1986, on the subject of unionizing IBM.
21. Michael Brody, op. cit., p. 90.
22. Richard Sennett and Jonathan Cobb, *The Hidden Injuries of Class* (New York, Vintage Books, 1973).
23. From the AFL-CIO Research Department, quoted in *Economic Notes* 53, no. 10 (October 1985), p. 2.
24. Federal Reserve Board study, quoted in *Economic Notes* 53, no. 9 (September 1985), p. 3.
25. Audrey Freedman, Conference Board report, March 1985, quoted in *Economic Notes* 53, no. 9 (September 1985), p. 1.
26. *Economic Notes* 53, no. 9 (September 1985) p. 2.
27. CRS study of fringe benefits, quoted in *Economic Notes* 53, no. 9 (September 1985), p. 3.
28. Meeting with Rochell, January 27, 1986.
29. AFL-CIO Committee on the Evolution of Work, op. cit., p. 11.
30. AFL-CIO Committee on the Evolution of Work, op. cit., p. 13.
31. Richard B. Freeman and James L. Medoff, *What Do Unions Do?* (New York: Basic Books, 1984), pp. 6–11.
32. Quoted in "Troubles of U.S. Labor Unions Eased in 1986," *The New York Times* (February 15, 1987), p. 31.

Chapter 6

1. An insightful overview of corporate training programs can be found in the Carnegie Foundation Special Report, *Corporate Classrooms: The Learning Business* by Nell P. Eurich (Princeton, N.J.: Carnegie Foundation for the Advancement of Teaching, 1985). *The World Labor Report 1: Employment, Incomes, Social Protection, New Information Technology* (Geneva, Switzerland: International Labour Organization, 1984) also discusses corporate training.

2. Ken Auletta, *The Underclass* (New York: Random House, 1982).

3. Ibid., p. xvi.

4. Gordon Berlin, cited in Paula Duggan, *Shaping the Work Force of the Future: An Agenda for Change* (Washington, D.C.: Northeast-Midwest Institute, 1984), p. 12.

5. Chicago Word Processing, described in "Basic Skills on the Job," *BCEL Newsletter for the Business Community* (January 1985), p. 6.

6. Sar A. Levitan and Clifford M. Johnson, *Beyond the Safety-Net* (Cambridge, Mass.: Ballinger, 1984), p. 36. The same argument is made in Greg Duncan's *Years of Poverty, Years of Plenty: The Changing Fortunes of American Workers and Families* (Ann Arbor: Survey Research Center, Institute for Social Research, University of Michigan, 1984).

7. Center for Public Resources, *Basic Skills in the U.S. Workforce*, 1983, summarized in Paula Duggan, *Literacy at Work: Developing Adult Basic Skills for Employment* (Washington, D.C.: Northeast-Midwest Institute, 1985), p. 5.

8. On illiteracy, see Jonathan Kozol, *Illiterate America* (Garden City, N.Y.: Anchor Press/Doubleday, 1985).

9. In Paula Duggan, op. cit., pp. 14–15.

10. Based on discussion in James M. Rosbron, "Unemployment Insurance Marks the 50th Anniversary," *Monthly Labor Review* 108, no. 9 (September 1985), pp. 21–28.

11. Eli Ginzberg, *Good Jobs, Bad Jobs, No Jobs* (Cambridge, Mass.: Harvard University Press, 1979), p. 155.

12. Ibid., p. 156.

13. Sar A. Levitan and Robert Taggart III, *Social Experimentation and Manpower Policy: The Rhetoric and the Reality* (Baltimore: Johns Hopkins University Press, 1971).

14. Gary Walker, Hilary Feldstein, and Katherine Solow, "An Independent Sector Assessment of the Job Training Partnership Act, Phase II: Initial Implementation, Executive Summary," Gunker Walker and Associates for the National Commission for Employment Policy, January 1985.

15. This discussion of the workforce is based on research conducted by C. Spellman for Abt Associates on the Food Stamp Work Registration and Job Search Demonstration Project (contract no. 53-3198-0-85, Food and Nutrition Service, U.S. Department of Agriculture).

16. Quoted in "Job Program Called a Failure," *The New York Times* (January 31, 1985), I, p. 7.

17. Eli Ginzberg, op. cit., the argument runs through chapters 10–17, pp. 105–178.

Chapter 7

1. National Commission on Jobs and Small Businesses, *Making America Work Again: Jobs, Small Business and the International Challenge* (Washington, D.C., 1987) p. 6.

2. Lester C. Thurow, *The Zero-Sum Solution: Building a World-Class American Economy* (New York: Simon and Schuster, 1985), ch. 1, "Where Are We?" pp. 27–44 and 48–49.
3. See Bruce Steinberg, "The Mass Market Is Splitting Apart," *Fortune* (November 28, 1983), pp. 76–82; also David Wessel, "Growing Gap: U.S. Rich and Poor Increase in Numbers: Middle Loses Ground," *Wall Street Journal* (September 22, 1986), p. 1.
4. Quoted in *Education Leadership* 64 (January 1985), p. 14.
5. Lester C. Thurow, op. cit., p. 67.
6. Lester C. Thurow, op. cit., p. 61. See also pp. 60–67 for a general discussion of the decline of the middle class.
7. The point about voter turnout is made by both Thurow, op. cit., and by George Gallup with William Proctor, *Forecast 2000: George Gallup, Jr., Predicts the Future of America* (New York: William Morrow, 1984).
8. Economist James Robinson, cited in Michael Harrington, *The New American Poverty* (New York: Holt, Rinehart and Winston, 1984).
9. Based on an interview at the New York City Branch of the American Institute of Banking, June 1986.
10. Quoted in Lester C. Thurow, op. cit., pp. 139–143.
11. The whole of the November 1985 issue was devoted to analysis of revised projections for the labor force in 1995.
12. George T. Selvestri and John M. Lukasiewicz, "Occupational Employment Projections: The 1984–95 Outlook," *Monthly Labor Review* 108, no. 11 (November 1985) p. 37.
13. Table 3 from "Occupations with the Largest Job Growth, 1984–95," *Monthly Labor Review* 108, no. 11 (November 1985), p. 51.
14. Valerie A. Personick, "A Second Look at Industry Output and Employment Trends to 1995," *Monthly Labor Review* 108, no. 11 (November 1985), p. 37.
15. Table 5 from "Fastest Declining Occupations," *Monthly Labor Review* 108, no. 11 (November 1985), p. 53.
16. Personick, op. cit., p. 52.
17. Ibid., p. 56.
18. Bureau of Labor Statistics data quoted by Peter Nulty in "Pushed Out at 45— Now What?" *Fortune* 115, no. 5 (March 2, 1987), p. 26.
19. Letter, December 20, 1985.
20. Ibid.
21. Letter, November 5, 1985.
22. Data on work days is from Thurow, op. cit., p. 139.

Chapter 8

1. Barry Bluestone and Bennett Harrison, *The Deindustrialization of America: Plant Closings, Community Abandonment and the Dismantling of Basic Industry* (New York: Basic Books, 1982).
2. Quoted by Robert Kuttner in "The Declining Middle," *Atlantic Monthly* (July 1983), p. 60.

3. Eileen Applebaum, *Technology and the Design of Jobs in the Insurance Industry* (Stanford, Calif.: Institute for Research on Educational Finance and Governance, 1984), quoted in U.S. Department of Labor, *Women and Office Automation: Issues for the Decade Ahead* (Washington, D.C.: U.S. Government Printing Office, 1985), p. 11.

4. Bureau of Labor Statistics projections presented in "The Myth of High Tech Jobs," *Harpers* (August 1984), p. 24.

5. Diana Rouse, research director, "Crisis and Computerization: The Uncertain Future of Clerical Work in the Great Lakes States." Final report submitted to the Joyce Foundation by the Working Women Education Fund, May 15, 1985.

6. David J. Roessner, project director, "Impact of Office Automation on Office Workers: A Technical Summary," draft document prepared for the U.S. Department of Labor, Employment and Training Administration (Atlanta, Ga: Georgia Institute of Technology, April 1984) p. 13.

7. Personal interview, fall 1985.

8. For a cogent discussion of farming out work see the U.S. Department of Labor, op. cit.

9. For a more in-depth discussion of the ties that anchor the blue-collar labor force, see Barry Bluestone and Bennett Harrison, op. cit.

10. Most of the evidence for this point has come from reading about towns where the migration of jobs has been extreme. The particular case I am thinking of came from discussions with administrators of a job training demonstration project in the Buffalo area in 1984.

11. U.S. Department of Labor, op. cit.

12. *The New York Times* (February 7, 1986), p. 15.

13. Neal H. Rosenthal, "The Shrinking Middle Class: Myth or Reality?" *Monthly Labor Review* 108, no. 3 (March 1985), pp. 9–10.

14. Letter from Secretary Brock, November 5, 1985. See also the "Afterword."

15. Based on Bruce Steinberg, "The Mass Market Is Splitting Apart," *Fortune* (November 28, 1983), pp. 76–82.

16. Quoted in ibid., p. 81.

17. Ibid., p. 82.

18. David Wessel, "Growing Gap: U.S. Rich and Poor Increase in Numbers: Middle Loses Ground," *Wall Street Journal* (September 22, 1986), p. 1.

Conclusion

1. Theodore Roszak, *Cult of Information: The Folklore of Computers and the True Art of Thinking* (New York: Pantheon, 1986); and Harley Shaiken, *Work Transformed: Automation and Labor in the Computer Age* (New York: Holt, Rinehart and Winston, 1984).

Bibliography

Abt, Clark C., ed. *Problems in American Social Policy Research*. Cambridge, Massachusetts: Abt Books, for the Council for Applied Social Research, 1980.

Adkins, Nelson F., ed. *Thomas Paine: Common Sense and Other Political Writings*. New York: The Liberal Arts Press, 1953.

AFL-CIO Committee on the Evolution of Work. "The Changing Situation of Workers and Their Unions: A Report on the Future of Work." Washington, D.C.; AFL-CIO, August 1985.

"America Rushes to High Technology Growth." *Business Week*, March 28, 1983, pp. 84–90.

American Management Association, Don Lee Bohl, ed. "The 1985 AMA Report on Information Centers." New York: American Management Association, 1985.

Anderson, Nels. *The Right to Work*. New York: Modern Age Books, 1938.

Anderson, Regina B. "Prosperity Projected for New Jersey-New York-Connecticut Urban Region, Keeping Up with U.S. Growth by 1990—If Housing Is Built for Employees." Attachment to Regional Plan Association Press Release no. 1538 (release date August 12, 1985).

Angarten, Stan. *Bit by Bit: An Illustrated History of Computers*. New York: Ticknor and Fields, 1984.

Applebaum, Eileen. *Technology and the Design of Jobs in the Insurance Industry*. Stanford, California: Institute for Research on Educational Finance and Governance, 1984.

Aronowitz, Stanley, Diana Fenton, and Bill Kornblum. "Preliminary Findings and Recommendations Regarding the Possible Effects of Computer-Aided Design on Members of Local 375 AFSME, District Council 37." New York: Institute on Labor and Community, May 18, 1984.

Ashley, Steven. "Artificial Intelligence in the Plant." *American Metal Market/Metalworking News*. Janaury 10, 1983, pp. 7–12.

Auletta, Ken. *Hard Feelings: Reporting on Politics, the Press, People and the City*. New York: Random House, 1980.

Auletta, Ken. *The Underclass*. New York: Random House, 1982.

Bartsch, Charles, with Karen Bachler, and Peter H. Doyle, project consultants. *Reaching for Recovery: New Economic Initiative in Michigan*. Washington, D.C.: Northeast-Midwest Institute/The Center for Policy Research, 1985.

Bell, Daniel. *The Coming of Post-Industrial Society: A Venture in Social Forecasting*. New York: Basic Books, 1973.

Bell, Daniel. "The Social Framework of the Information Society." In *The Microelectronics Revolution* 3rd ed., Tom Forester, ed. Cambridge, Massachusetts: MIT Press, 1983, pp. 500–549.

Bell, Daniel. "Teletext and Technology: The New Networks of Knowledge and Information in Post-Industrial Society." *Encounter* 48, no. 6, June 1977, pp. 9–29.

Bell, Daniel, ed. *Toward the Year 2000: Work in Progress*. Boston, Massachusetts: Houghton Mifflin, 1968. (American Academy of Arts and Sciences, 1967, summer issue of *Daedalus*).

Bellah, Robert N., Richard Madsen, William M. Sullivan, Ann Sindler, and Steven M. Typson. *Habits of the Heart: Individualism and Commitment in American Life*. Berkeley, California: University of California Press, 1985.

Berg, Maxine. *The Machinery Question and the Making of Political Economy 1815–1848*. London: Cambridge University Press, 1980.

Berlin, Gordon and Joanne Duhl. *Education, Equity and Economic Excellence: The Critical Role of Second Chance Basic Skills and Job Training Programs*. August 30, 1984.

Blauner, Robert. *Alienation and Freedom: The Factory Worker and His Industry*. Chicago: The University of Chicago Press, 1964.

Bluestone, Barry and Bennett Harrison. *The Deindustrialization of America: Plant Closings, Community Abandonment and the Dismantling of Basic Industry*. New York: Basic Books, 1982.

Bolter, David J. *Turing's Man: Western Culture in the Computer Age*. Chapel Hill, North Carolina: University of North Carolina Press, 1984.

Boorstin, Daniel J. *The Republic of Technology: Reflections on Our Future Community*. New York: Harper and Row, 1978.

Borus, Michael E. and William R. Tash. *Measuring the Impact of Manpower Programs: Policy Papers in Human Resources and Industrial Relations 17*. Ann Arbor, Michigan: Institute of Labor and Industrial Relations, the University of Michigan-Wayne State University, 1970.

Botsch, Robert Emil. *We Shall Not Overcome: Populism and Southern Blue-Collar Workers*. Chapel Hill, North Carolina: The University of North Carolina Press, 1980.

Boyle, Peter H. and Candice Brisson. *Partners in Growth: Business-Higher Education Development Strategies*. Washington, D.C.: Northeast-Midwest Institute, Center for Regional Policy-Education-Economic Development Series, 1985.

Brave New World? Living with Information Technology. Oxford, England: Pergamon Press, 1983.

Braverman, Harry. "The Degradation of Work in the Twentieth Century." *Monthly Review* 1, no. 34, May 1982, pp. 1–13.

Braverman, Harry. *Labor and Monopoly Capital*. New York: Monthly Review Press, 1974.

Brecher, Jeremy and David Montgomery. "Crisis Economy: Born-Again Labor Movement." *Monthly Review* 35, no. 10, March 1984, pp. 1–23.

Brich, David. "Matters of Fact," interviewed in *Inc.*, April 1985, pp. 31–36.

Brody, Michael. "Meet Today's Young American Worker." *Fortune*, November 11, 1985.

Buick, Gilbert. "The Boundless Age of the Computer." *Fortune* 69, no. 3, March 1964, pp. 101–111, 230–232.

Chamot, Denis, ed. "Technological Change and Unions: An International Perspective." AFL-CIO Department of Professional Employees, Publication no. 82-2.

Clawson, Dan. *Bureaucracy and the Labor Process: The Transformation of U.S. Industry, 1860–1920.* New York: Monthly Review Press, 1980.

Clegg, Stewart. "Organization and Control." *Administrative Science Quarterly* 26, 1981, pp. 545–562.

Communication Workers of America. *Committee on the Future Report.* Washington, D.C.: Communication Workers of America, March 1983.

"Computers and Electronics." Special issue of *Science* 215, no. 4534, February 12, 1982.

Congressional Research America. *The Computer Revolution and the U.S. Labor Force.* Washington, D.C.: U.S. Government Printing Office, 1985.

Cooley, Mike. *Architect or Bee? The Human/Technology Relationship.* Boston, Massachusetts: South End Press, 1980.

Crittenden, Ann. "The Age of 'Me-First' Management." *The New York Times*, business section, August 19, 1984.

David, Henry. *Manpower Policies for a Democratic Society: The Final Statement of the Council* [National Manpower Council]. New York: Columbia University Press, 1965.

Dechert, Charles R., ed. *The Social Impact of Cybernetics.* New York: Simon and Schuster, 1966.

Deken, Joseph. *The Electronic Cottage.* New York: William Morrow, 1982.

Dertouzos, Michael, ed. *The Computer Age: A Twenty Year View.* Boston, Massachusetts: MIT Press, 1979.

Dizard, Wilson P. *The Coming Information Age: An Overview of Technology, Economics, and Politics.* New York: Longman, 1982.

Draper, Roger "The Golden Arm." *New York Review of Books*, October 24 1985, pp. 46–52

Drennan, Matthew P. *Implications of Computer and Communication Technology for Less Skilled Service Employment Opportunities.* Final Report to U.S. Department of Labor (under grant #USDL 21-36-80-31). New York: Columbia University Conservation of Human Resources, January 21, 1983.

Dreyfus, Hubert. *What Computers Can't Do: A Critique of Artificial Reason.* New York: Harper and Row, 1972.

Dreyfus, Hubert. "What Expert Systems Can't Do." *Raritan*, Spring 1984, pp. 22–36.

Drucker, Peter. "Our Entrepreneurial Economy." *Harvard Business Review* 1, January–February 1984, pp. 59–64.

DuBois, Thomas, staff researcher, Midwest Center for Labor Research. *The Call That Never Came: The Return of Steel Production without the Return of Steel Workers in Northwest Indiana, 1979–1984*. Chicago: Calumet Project for Industrial Jobs (a joint effort of the Midwest Center for Labor Research and United Citizens Organization), 1985.

DuBois, Thomas, with David Bensman, Greg LeRoy, and Jack Metzgar. *Where Have All the Jobs Gone? Causes of Job Loss Past and Future in the Northwest Indiana Steel Industry*. Chicago: Calumet Project for Industrial Jobs (a joint effort of the Midwest Center for Labor Research and United Citizens Organization), 1985.

Duggan, Paula. *Literacy at Work: Developing Adult Basic Skills for Employment*. Washington, D.C.: Northeast-Midwest Institute, 1985.

Duggan, Paula. *Shaping the Work Force of the Future: An Agenda for Change*. Washington, D.C.: Northeast-Midwest Institute, 1984.

Duggan, Paula, and Virginia Meyer. *The New American Unemployment: Appropriate Government Responses to Structural Dislocation*. Washington, D.C.: Northeast-Midwest Institute, Center for Regional Policy, 1985.

Duncan, Greg J. *Years of Poverty, Years of Plenty: The Changing Economic Fortunes of American Workers and Families*. Ann Arbor, Michigan: Survey Research Center, Institute for Social Research, University of Michigan, 1984.

Edwards, Richard. *Contested Terrain: The Transformation of the Workplace in the Twentieth Century*. New York: Basic Books, 1979.

Estey, Marten. *The Unions: Structure, Development and Management*. New York: Harcourt Brace Jovanovich, 1981.

Etzioni, Amitar. *An Immodest Agenda: Rebuilding America before the Twenty-First Century*. New York: McGraw-Hill, 1983.

Eurich, Nell P., with a foreword by Ernest L. Boyer, *Corporate Classrooms: The Learning Business*. A Carnegie Foundation Special Report. Princeton, New Jersey: Carnegie Foundation for the Advancement of Teaching, 1985.

Evans, Christopher Riche. *The Making of the Micro: A History of the Computer*. New York: Van Nostrand Reinhold Company, 1981.

Firschein, Oscar, et al. "Forecasting and Assessing the Impact of Artificial Intelligence on Society." Proceedings of the International Joint Conference of Artificial Intelligence, Palo Alto, California: Stanford University, August 20–23, 1973.

Fisher, Anne B. "Ford Is Back on Track." *Fortune* 14, December 23, 1985.

Fishman, Katherine Davis. *The Computer Establishment*. New York: McGraw-Hill, 1981.

Flaim, Paul O. and Ellen Siligal. "Displaced Workers of 1979–83: How Well Have They Fared?" *Monthly Labor Review*, June 19, 1985.

Flint, Jerry. "We're Going To Do Just Fine." *Forbes*, October 21, 1985, pp. 37–40.

Forester, Tom, ed. *The Microelectronics Revolution: The Complete Guide to the New Technology and Its Impact on Society*. Cambridge, Massachusetts: MIT Press, 1981.

Fossum, John A. *Labor Relations: Development, Structure, Process* 3rd ed. Plains, Texas: Business Publications, 1979, 1985.

Freeman, Richard B. and James L. Medoff. *What Do Unions Do?* New York: Basic Books, 1984.

Gallup, George Jr. with William Proctor. *Forecast 2000: George Gallup, Jr., Predicts the Future of America*. New York: William Morrow, 1984.

Garranty, John A. *Unemployment in History: Economic Thought and Public Policy*. New York: Harper and Row Colophon Books, 1978.

Gevarter, William B. *An Overview of Artificial Intelligence and Robotics*. National Bureau of Standards Document no. NBSIR 82-2479. Washington, D.C.: Department of Commerce, March 1982.

Gevarter, William B. *An Overview of Expert Systems*. National Bureau of Standards Document no. NBSIR 82-2505. Washington, D.C.: Department of Commerce, May 1982.

Gibbons, John H, director, Office of Technology Assessment. *Automation of America's Offices, 1985–2000*. OTA-CIT-287. Washington, D.C.: U.S. Government Printing Office, December, 1985.

Giddens, Anthony and David Held, eds. *Classes, Power, and Conflict: Classical and Contemporary Debates*. Berkeley, California: University of California Press, 1982.

Ginzberg, Eli. *Good Jobs, Bad Jobs, No Jobs*. Cambridge, Massachusetts: Harvard University Press, 1979.

Ginzberg, Eli. "The Mechanization of Work." *Scientific American* 247, no. 3, September 1982, pp. 67–164.

Giuliano, Vincent E. "The Mechanization of Office Work." *Scientific American* 247, no. 3, September 1982, pp. 149–164.

Goldhaber, Michael. "Politics and Technology: Microprocessors and the Present Prospect of a New Industrial Revolution." *Socialist Review* 52, vol. 10, no. 4, July–August 1980, pp. 9–32.

Gordon, David M., Richard Edwards, and Michael Reich. *Segmented Work, Divided Workers: The Historical Transformation of Labor in the United States*. Cambridge, England: Cambridge University Press, 1982.

Gorz, Andre. *Farewell to the Working Class: An Essay on Post-Industrial Socialism*. Boston, Massachusetts: South End Press, 1982.

Gorz, Andre. *Strategy for Labor: A Radical Proposal*, trans. Martin A. Nicholaus and Victoria Ortiz. Boston, Massachusetts: Beacon Press, 1967.

Gould, Jay M. *The Technical Elite*. New York: Augustus M. Kelley, 1966.

Graham, Neill. *The Mind Tool: Computers and Their Impact on Society*. St. Paul, Minnesota: West, 1980.

Green, James R. *The World of the Worker: Labor in Twentieth-Century America*. New York: Hill and Wang, 1980.

Harrington, Michael. *The New American Poverty*. New York: Holt, Rinehart and Winston, 1984.

Harrington, Michael. *The Other America: Poverty in the United States*. New York: Macmillan, 1962.

Herman, Edward S. *Corporate Control; Corporate Power*. Cambridge, England: Cambridge University Press, 1981.

Hill, Richard. "The Coming of Post Industrial Society." *The Insurgent Sociologist* 4, no. 3, Spring 1974, pp. 37–51.

Hill, Stephen. *Competition and Control at Work*. Cambridge, Massachusetts: MIT Press, 1981.

Hindle, Brooke. *Technology in Early America: Needs and Opportunities for Study.* Chapel Hill, North Carolina: The University of North Carolina Press, 1966.

Hirschhorn, Larry. *Beyond Mechanization: Work and Technology in the Post-Industrial Age.* Cambridge, Massachusetts: MIT Press, 1984.

Hirschhorn, Larry. "The Post-Industrial Labor Process." *New Political Science* 7, vol. 2, no. 3, Fall 1981, pp. 11–32.

Hobsbawm, Eric J. "The Machine Breakers." *Past and Present* 1, 1970, pp. 52–70.

Howard, Robert. *Brave New Workplace.* New York: Viking Press, 1984.

Howe, Charles L. "Big Labor and Big Blue." *Datamation*, January 1, 1986.

Industrial Democracy in Europe International Research Group. *European Industrial Relations.* Oxford, England: Clarendon Press, 1981.

Institute for Labor Education and Research. *What's Wrong with the U.S. Economy? A Popular Guide for the Rest of Us.* Boston, Massachusetts: South End Press, 1982.

International Association of Machinists and Aerospace Workers, William W. Winpisinger, international president, and Eugene Glover, secretary-treasurer. *Let's Rebuild America, with the Text of the IAM's Rebuilding America Act, Including the Technology Bill of Rights.* Washington, D.C.: The International Association of Machinists and Aerospace Workers, 1983.

International Labour Organization. *World Labor Report 1: Employment Incomes, Social Protection, New Information Technology.* Geneva, Switzerland: International Labour Organization, 1984.

Kahn, Herman. *The Coming Boom: Economic, Political and Social.* New York: Simon and Schuster, 1982.

Kaplinsky, Raphael. *Automation: The Technology and Society.* Harlow, England: Longman, 1984.

Katzell, Raymond A.. and Daniel Yankelovitch. *Work, Productivity and Job Satisfaction: An Education of Policy-Related Research.* New York: The Psychological Corporation, 1975.

Kerr, Clark. *The Future of Industrial Societies: Convergence or Continuing Diversity.* Cambridge, Massachusetts: Harvard University Press, 1983.

Kidder, Tracy. *The Soul of a New Machine.* Boston, Massachusetts: An Atlantic Monthly Press Book; Little, Brown, 1981.

Kiesler, Sara. "Computers and Information Distribution in Organizations Or, How Computers Will Change Who Knows What." Prepared for Symposium, Ancient Humans in Tomorrow's Electronic World, Aspen Institute for Humanistic Studies, 1984.

Kozol, Jonathan. *Illiterate America.* Garden City, New York: Anchor Press/Doubleday, 1985.

Kraft, Philip. *Programmers and Managers: The Routinization of Computer Programming in the United States.* New York: Springer-Verlag, 1977.

Kuttner, Robert. "The Declining Middle." *Atlantic Monthly* 252, July 1983, pp. 60–72.

Kuttner, Robert. "Jobs." *Dissent* 31, no. 1, winter 1984, pp. 30–41.

"The Labor Process." Special issue of *The Insurgent Sociologist* 11, no. 3, fall 1982.

Laver, Murray. *Computers and Social Change.* Cambridge, Massachusetts: Cambridge University Press, 1980.

Lawrence, Robert Z. *Can America Compete?* Washington, D.C.: The Brookings Institution, 1984.

Leff, Wall and Marilyn Haft, eds. *Time without Work.* Boston, Massachusetts: South End Press, 1983.

Leontief, Wassily W. "The Distribution of Work and Income." *Scientific American* 247, no. 3, September 1982, pp. 188–204.

Leontief, Wassily W. "The Long-Term Impact of Technology on Employment and Unemployment." Speech delivered at the National Academy of Engineering Symposium, June 30, 1983.

Leontief, Wassily W. and Faye Duchin. *The Impact of Automation on Employment, 1963–2000.* Final report to the National Science Foundation. New York: Institute for Economic Analysis, New York University, 1984.

Leventman, Paula G. *Professionals Out of Work.* New York: The Free Press, 1981.

Levitan, Sar A. and Clifford M. Johnson. *Beyond the Safety-Net.* Cambridge, Massachusetts: Ballinger, 1984.

Levitan, Sar A. and Clifford M. Johnson. "The Future of Work: Does It Belong to Us or to the Robots?" *Monthly Labor Review*, September 1982.

Levitan, Sar A. and Clifford M. Johnson. *Second Thoughts on Work.* Kalamazoo, Michigan: W.E. Upjohn Institute for Employment Research, 1982.

Levitan, Sar A. and Robert Taggart III. *Social Experimentation and Manpower Policy: The Rhetoric and the Reality.* Baltimore, Maryland: The Johns Hopkins Press, 1971.

Lund, Leonard. "Business in the Community: Where Vanguard Companies Are Focusing." The Conference Board Research Bulletin no. 185. New York: The Conference Board, 1985.

Lusterman, Seymour. "Trends in Corporate Education and Training." The Conference Board Report no. 870. New York: The Conference Board, 1985.

McCorduck, Pamela. *Machines Who Think: A Personal Inquiry into the History and Prospects of Artificial Intelligence.* San Francisco: W.H. Freeman, 1979.

Magnet, Myron. "Restructuring Really Works." *Fortune* 115, March 2, 1987, pp. 38–48.

Malcolm, Andrew H., "What Five Families Did after Losing the Farm." *The New York Times*, February 4, 1987, pp. A1, A18.

Martin, James. *Telematic Society—A Challenge for Tomorrow.* Englewood Cliffs, New Jersey: Prentice-Hall, 1981.

Martin, James and Adrian R.D. Norman. *The Computerization of Society: An Appraisal of the Impact of Computers on Society over the Next Fifteen Years.* Harmondsworth, England: Penguin, 1973.

Masuda, Yonegi. *The Information Society as Post-Industrial Society.* Tokyo: Institute for the Information Society, 1981.

Mayer, Martin. *The Bankers.* New York: Weybright and Talley, 1974.

Mayer, Martin. *The Money Bazaar: Understanding the Banking Revolution around Us.* New York: E.P. Dalton, 1984.

"The Mechanization of Work." Special Issue of *Scientific American* 247, no. 3, September 1982.

Meier, Gerald M. *Emerging from Poverty: The Economics That Really Matter.* New York: Oxford University Press, 1984.

Mollenkopf, John. "A Strategic Approach to Education for Economic Development in New York City." CUNY Graduate Center, Political Science Program, draft, October 2, 1984.

Morison, Elting E. *Men, Machines and Modern Times*. Cambridge, Massachusetts: MIT Press, 1966.

Morris, R.J. *Class and Class Consciousness in the Industrial Revolution 1780–1850*. London: Macmillan Press, 1979.

Moss, Mitchell L. "The Telecommunications Infrastructure in the City of New York." New York University, submitted to Division of Policy Analysis Office for Economic Development, City of New York, 1985.

Mumford, Lewis. *Technics and Civilization*. New York: Harcourt Brace, 1934.

Naisbitt, John. *Megatrends: Ten New Directions Transforming Our Lives*. New York: Warner Books, 1982, 1984.

Naisbitt, John and Patricia Aburdene. *Re-Inventing the Corporation: Transforming Your Job and Your Company for the New Information Society*. New York: Warner Books, 1985.

National Commission on Jobs and Small Businesses. *Making America Work Again: Jobs, Small Business and the International Challenge*. Washington, D.C.: National Commission on Jobs and Small Businesses, 1987.

"A New Era for Management." Special report in *Business Week*, April 25, 1984, pp. 50–86.

Newman, Edwin, reporter. Transcript of *NBC: Marvelous Machines. . . Expendable People*. New York: National Broadcasting Company, 1983.

Neilsen, John. "Management Layoffs Won't Quit." *Fortune*, October 28, 1985, pp. 46–50.

Noble, David F. *America by Design: Science, Technology, and the Rise of Corporate Capitalism*. Oxford, England: Oxford University Press, 1977.

Noble, David F. *Forces of Production: A Social History of Industrial Automation*. New York: Knopf, 1984.

Noble, David F. "Present Tense Technology (Part I)." *Democracy* 3, no. 2, Spring 1983, pp. 8–27.

Noble, David F. "Present Tense Technology (Part II)." *Democracy* 3, no. 3, summer 1983, pp. 70–82.

Noble, David F. "Present Tense Technology (Part III)." *Democracy* 3, no. 4, fall 1983, pp. 71–93.

Noble, Douglas. "The Underside of Computer Literacy." *Raritan*, spring 1984, pp. 37–64.

Noble, Kenneth B. "Labor Department Data Show a Revival of Major Strikes." *The New York Times*, September 21, 1986.

Noble, Kenneth B. "Steel Union Locals Back USX Pact." *The New York Times*, January 19, 1987.

Nora, Simon and Alain Minc. *The Computerization of Society: A Report to the President of France*. Cambridge, Massachusetts: MIT Press, 1981.

Norman, Colin. "The New Industrial Revolution: How Microelectronics May Change the Workplace." In *Working: Changes and Choices*, James O'Toole, et al., eds. New York: Human Sciences Press, 1981.

Noyelle, Thierry J. "Services, Urban Economic Development and Industrial Policy: Some Central Linkages." Paper presented at symposium on The Industrial Policy Question: State and Local Issues, May 17 and 18, 1984, University of North Carolina, Chapel Hill.

Nulty, Peter. "Pushed Out at 45—Now What?" *Fortune* 115, no. 5, March 2, 1987, pp. 26–30.

Olmsted, Barney. "Job-Sharing—A New Way to Work." In *Working: Changes and Choices*, James O'Toole, et al., eds. New York: Human Sciences Press, 1981, pp. 325–330.

O'Toole, James. "The Reserve Army of the Underemployed." *Change*, May 1975, pp. 26–63.

Ouchi, William G. *The M-Form Society: How American Teamwork Can Recapture the Competitive Edge*. Reading, Massachusetts: Addison-Wesley, 1984.

Pacey, Arnold. *The Culture of Technology*. Cambridge, Massachusetts: MIT Press, 1983.

Palmer, John C., ed. *Creating Jobs: Public Employment Programs and Wage Subsidies*. Washington, D.C.: The Brookings Institution, 1978.

Patterson, James T. *America's Struggle against Poverty 1900–1980*. Cambridge, Massachusetts: Harvard University Press, 1981.

Peccei, Aurelio. *One Hundred Pages for the Future: Reflections of the President of the Club of Rome*. New York: Pergamon Press, 1981.

Personick, Valerie A. "A Second Look at Industry Output and Employment Trends to 1995." *Monthly Labor Review* 108, no. 11, November 1985.

Peters, Thomas J. and Nancy Austin. *A Passion for Excellence: The Leadership Difference*. New York: Random House, 1985.

Peters, Thomas J. and Robert H. Waterman Jr. *In Search of Excellence*. New York: Harper and Row, 1982.

Reich, Robert B. *The Next American Frontier*. New York: Times Books, 1983.

Riesman, David. "Leisure and Work in Post-Industrial Society." In *Abundance for What? and Other Essays*. Garden City, New York: Doubleday, 1964, pp. 162–183.

"Productivity and Displacement Effects." *Science* 195, no. 4283, March 18, 1977, pp. 1179–1184.

Rochell, Carlton C., ed. *An Information Agenda for the 1980s: Proceedings of a Colloquium June 17–18, 1980*. Chicago: American Library Association, 1981.

Rodgers, Daniel T. "Work Ideals and the Industrial Invasion." In *Working: Changes and Choices*. James O'Toole, et al., eds. New York: Human Sciences Press, 1981, pp. 134–144.

Roessner, David J. "Impact of Office Automation on Office Workers: A Technical Summary," draft report prepared for the U.S. Department of Labor, Employment and Training Administration. Atlanta, Georgia: Georgia Institute of Technology, April 1984.

Rohatyn, Felix G. *The Twenty-Year Century: Essays on Economics and Public Finance*. New York: Random House, 1983.

Rosbron, James M. "Unemployment Insurance Marks the 50th Anniversary." *Monthly Labor Review* 108, no. 9, September 1985.

Rosenthal, Neal H. "The Shrinking Middle Class: Myth or Reality?" *Monthly Labor Review* 108, no. 3, March 1985.

Rostow, W.W. *The Barbaric Counter-Revolution: Cause and Cure.* Austin, Texas: University of Texas Press, 1983.

Roszak, Theodore. *Cult of Information: The Folklore of Computers and the True Art of Thinking.* New York: Pantheon, 1986.

Rouse, Diana, research director. "Crisis and Computerization: The Uncertain Future of Clerical Work in the Great Lakes States." Final report submitted to the Joyce Foundation by the Working Women Education Fund, May 15, 1985.

Rukeyser, Louis. *What's Ahead for the Economy: The Challenge and the Chance.* New York: Simon and Schuster, 1983.

Samuelson, Robert J. "Our Computerized Society." *Newsweek*, September 9, 1985, p. 73.

Sawhill, Isabel V. "Reaganomics in Retrospect." In *Perspectives on the Reagan Years*, John L. Palmer, ed. Washington, D.C.: The Urban Institute Press, 1986.

Schlozman, Kay Lehman and Sidney Verba. *Injury to Insult: Unemployment, Class and Political Response.* Cambridge, Massachusetts: Harvard University Press, 1979.

Selvestri, George T. and John M. Lukasiewicz. "Occupational Employment Projections: The 1984–95 Outlook." *Monthly Labor Review* 108, no. 11, November 1985.

Sennett, Richard and Jonathan Cobb. *The Hidden Injuries of Class.* New York: Vintage Books, 1973.

Shaiken, Harley. *Work Transformed: Automation and Labor in the Computer Age.* New York: Holt, Rinehart and Winston, 1984.

Shurkin, Joel. *Engines of the Mind: A History of the Computer.* New York: W.W. Norton, 1984.

Siegel, Lenny and John Markoff. *The High Cost of High Tech: The Darkside of the Chip.* New York: Harper and Row, 1985.

Simmons, John and William Mares. *Working Together: Employee Participation in Action.* New York: New York University Press, 1985.

Simon, Herbert. "What Computers Mean for Man and Society." In *The Microelectronics Revolution*, Tom Forester, ed. Oxford, England: Basil Blackwell, 1980.

Smith, Linda C. "Artificial Intelligence Applications in Information Systems." *Annual Review of Information Science and Technology* 15, 1980, pp. 67–105.

Stark, David. "Class Struggle and the Labor Process." *Theory and Society* 9, no. 1, January 1980, pp. 89–130.

Steinberg, Bruce. "The Mass Market Is Splitting Apart." *Fortune*, November 28, 1983, pp. 76–82.

Stokley, Richard. "The Tales of Four Hunters." *Fortune* 115, no. 5, March 2, 1987, pp. 31–37.

Szolovits, Peter, ed. *Artificial Intelligence in Medicine.* AAAS Selected Symposium. Boulder, Colorado: Westview Press, 1982.

"Technology and Jobs: Office Automation." *Occupational Outlook Quarterly*, spring 1985.

"Technology, the Labor Process, and the Working Class." Special issue of *Monthly Review* 28, no. 3, July–August 1976.

Terkel, Studs. *Working: People Talking about What They Do All Day and How They Feel about What They Do.* New York: Pantheon Books, 1974.

Theobald, Robert. "Cybernetics and the Problems of Social Organization." In *The Social Impact of Cybernetics*, C.R. Dechert, ed. New York: Simon and Schuster, 1966.

Therborn, Goran. "The Prospect of Labor and the Transformation of Advanced Capitalism." *New Left Review* 145, May–June 1984, pp. 5–38.

Thomis, Malcolm I. *The Luddites.* New York: Schocken Books, 1970.

Thompson, Edward P. "The Crime of Anonymity." In *Albion's Fatal Tree: Crime and Society in Eighteenth-Century England*, Douglas Hay, et al., eds. New York: Pantheon Books, 1975, pp. 255–344.

Thompson, Edward P. *The Making of the English Working Class.* New York: Vintage Books, 1963.

Thompson, Edward P. "Time, Work Discipline and Industrial Capitalism." *Past and Present* 38, 1967, pp. 56–97.

Thurow, Lester C. *The Zero-Sum Solution: Building a World-Class American Economy.* New York: Simon and Schuster, 1985.

Toffler, Alvin. *Previews and Premises.* New York: William Morrow, 1983.

Toffler, Alvin. *The Third Wave.* New York: William Morrow, 1980.

Toller, Ernst. *The Machine Wreckers.* New York: Alfred A. Knopf, 1923.

Touraine, Alain. *The Post-Industrial Society*, trans. Leonard F.X. Mayhew. New York: Random House, 1971.

Turkle, Sherry. *The Second Self: Computers and the Human Spirit.* New York: Simon and Schuster, 1984.

U.S. Department of Health, Education and Welfare. *Work in America.* Washington, D.C.: U.S. Government Printing Office, 1972.

U.S. Department of Labor. *Women and Office Automation: Issues for the Decade Ahead.* Washington, D.C.: U.S. Government Printing Office, 1985.

Unterweger, Peter. "Work, Automation and the Economy." Paper presented at the American Association for the Advancement of Science, annual meeting, Detroit, Michigan, May 27, 1983.

Walker, Gary, Hilary Feldstein, and Katherine Solow. "An Independent Sector Assessment of the Job Training Partnership Act. Phase II: Initial Implementation." Gunker Walker and Associates for the National Commission for Employment Policy, January 1985.

Wayne, Leslie. "Implications of USX Pact." *The New York Times*, January 22, 1987.

Weidenbaum, Murray C. *The Future of Business Regulation: Private Action and Public Demand.* New York: American Management Association Communications, 1979.

Weizenbaum, Joseph. *Computer Power and Human Reason.* San Francisco: Freeman, 1976.

Werneke, Diane. *Microelectronics and Working Women: A Literature Summary.* Washington, D.C.: National Academy Press for the Committee on Women's Employment and Related Issues, National Research Council, 1984.

Wessel, David. "Growing Gap: U.S. Rich and Poor Increase in Numbers: Middle Loses Ground." *Wall Street Journal*, September 22, 1986, p. 1.

Wiener, Norbert. *The Human Use of Human Beings* 2nd ed. Garden City, New York: Doubleday Anchor Books, 1950.

Williams, Raymond. *The Year 2000: A Radical Look at the Future—and What We Can Do to Change It*. New York: Pantheon Books, 1983.

Williams, Rosalind. "The Machine Breakers." *Technology Illustrated*, July 1983.

Winner, Langdon. "Do Artifacts Have Politics?" *Daedalus* 109, no. 1., winter 1980, pp. 121–136.

Winston, Patrick H. and Karen A. Prendergast, eds. *The AI Business: Commercial Uses of Artificial Intelligence*. Cambridge, Massachusetts: MIT Press, 1984.

Woodward, Joan. "Management and Technology." In *Organizational Theory*, D.S. Pugh, ed. London: Penguin Books, 1971.

Zimbalist, Andrew, ed. *Case Studies on the Labor Process*. New York: Monthly Review Press, 1979.

Zuboff, Shoshana. "New Worlds of Computer Mediated Work." *Harvard Business Review* 60, 5 September–October 1982, pp. 142–152.

Index

About the Authors

Carlton Rochell manages the libraries and the University Press at New York University. His education includes degrees in mathematics and economics, urban studies, and library and information science. He has spent several years totally automating the libraries of New York University and is currently directing a foundation-funded research project to develop a model for maximizing newer technologies for scholarly research.

Dr. Rochell is chairman of the board of the Research Libraries Group, a consortium of thirty-eight leading research universities. He is the author of scores of articles and two previous books: *Practical Administration of Public Libraries* (Harper, 1981) and *An Information Agenda for the 1980's* (American Library Association, 1981). A frequent speaker on the applications of technology and its benefits and costs, he has won numerous awards and honors for scholarship, for his professional work, and for his writing.

Christina Spellman teaches sociology at New York University and does independent research. She has undertaken extensive site analysis for the national Food Stamp Work Rules Demonstration Study, Analysis of Community Action Agencies in Appalachia, and, most recently, contributed research and analysis on the New York City waterfront for the City's Commission on the year 2000. Dr. Spellman lives in Manhattan.